Taxation of Big Data

Xavier Oberson • Alara Efsun Yazıcıoğlu

Taxation of Big Data

Springer

Xavier Oberson
Faculty of Law
University of Geneva
Geneva, Switzerland

Alara Efsun Yazıcıoğlu
Faculty of Law
Kadir Has University
Istanbul, Türkiye

ISBN 978-3-031-43371-9 ISBN 978-3-031-43372-6 (eBook)
https://doi.org/10.1007/978-3-031-43372-6

This Springer imprint is published by the registered company Springer Nature Switzerland AG
The registered company address is: Gewerbestrasse 11, 6330 Cham, Switzerland

Paper in this product is recyclable.

Dedications:
To Laure Mi Hyun Croset (Prof. Xavier Oberson)
To Gülbin, Yusuf, and Bisou Yazıcıoğlu (Assist. Prof. Alara Efsun Yazıcıoğlu)

Acknowledgments

This book was prepared with the support of a strong team, to whom the authors wish to express their gratitude. We would like to thank Dr. iur. Brigitte Reschke, the editor of this book, as well as Daniel Ignatius Jagadisan and Peter Pushpanathan, the project coordinators of this book, for their help and support from the very beginning to the very end. We would like to thank Agnes Mermoud, personal assistant of Prof. Oberson, for her support. We would like to thank Pascale Fontana and Marielle Martin-Sacristan, documentalists at Oberson Abels SA, for their assistance in the compilation of the relevant documentation throughout the writing process. We would like to thank Sıla Adak, teaching and research assistant at the Financial Law Section of Kadir Has University, for her outstanding work during the preparation of this manuscript, her valuable scientific contribution, and the continuous moral support she has provided us with throughout the entire process. We would like to thank Sena Bakay, the legal designer with whom we had the pleasure of collaborating during the preparation of all the figures used in this book, for her excellent designs, as well as her genuine enthusiasm for the subject.

We would like to express our deepest gratitude to our families who have, as usual, provided their infallible support, without which this manuscript could not have been completed. As an expression of our sincere gratefulness, this book is dedicated to them.

Contents

List of Abbreviations

AI	Artificial Intelligence
BEPS	Base Erosion and Profit Shifting
CJEU	Court of Justice of the European Union
DST	Digital services tax
DTT	Double tax treaty
EU	European Union
EUR	Euro
FE	Fixed establishment
GAFA	Google, Apple, Facebook and Amazon.com
GDP	Gross domestic product
GDPR	General Data Protection Regulation
GloBE	Global anti-Base Erosion
GPS	Global Positioning System
GST	Goods and services tax
IAS	International Accounting Standard
IMF	International Monetary Fund
IoT	Internet of Things
IP	Internet Protocol
ML	Machine learning
MLI	Multilateral Convention to Implement Tax Treaty Related Measures to Prevent Base Erosion and Profit Shifting (OECD)
MNE	Multinational enterprise
MTurk	Mechanical Turk (Amazon.com)
OECD	Organisation for Economic Co-operation and Development
OECD MC	OECD Model Tax Convention on Income and on Capital (2017)
OECD TP Guidelines	OECD Transfer Pricing Guidelines for Multinational Enterprises and Tax Administrations (2022)
PE	Permanent establishment
R&D	Research and development

ROT data	Redundant, Obsolete and Trivial data
SME	Small and medium-sized enterprise
SURE	Sustained user relationships
TP	Transfer pricing
UHRS	Universal Human Relevance System (Microsoft)
UK	United Kingdom
UN MC	United Nations Model Double Taxation Convention Between Developed and Developing Countries (2021)
UN	United Nations
US	United States
USD	United States dollar
VAT	Value added tax
WTO	World Trade Organisation

Chapter 1
Introduction

Traditionally considered as a mere component of the digital economy, data could only occupy a supporting role in the international tax law scene. The authors are convinced that analysing data on their own merits would reveal to be extremely beneficial not only from the perspective of data, but also from the general perspective of the digital economy. As a matter of fact, despite being neglected and constantly demoted to a supporting role, data are a leading actor of the digital economy. Not treating data with the pre-eminence that they deserve can only result in an unbalanced analysis and halfway solutions. Aiming to remedy to the situation, this book casts data, more specificially user data, as its leading actor. This implies providing data-specific solutions to data-driven problems.

First, some general information is provided about our leading actor. From their generation to their deletion, the life cycle of this new raw material and the youngest intangible is examined in Chap. 2. Then, the anatomy of 'the data business' is laid out. The main players active in the data ecosystem, the dynamics they currently give rise to as well as the main issues caused by those dynamics are examined in Chap. 3. Chapter 3 also establishes the backbone of the tax law analysis conducted in this book. It is the chapter where the four-step methodology is laid out. *Step one*, occurring in the generation and collection phases of the data's life cycle, comprises 'the digital barter' and other collection of data. *Step two*, taking place during the processing and analytics phases of the data's life cycle, consists of microwork. *Step three*, situated in the storage, processing, analytics and use phases of the data's life cycle, encompasses aggregation and internal use of data. *Step four*, materialising during the distribution and use phases of the data's life cycle, covers sale of data, transfer of data and granting the right to use a database.

The authors are well aware of the fact that tax law cannot possibly constitute a panacea to all economic problems. Quite the contrary, most issues are better solved not at the end of a chain but in its previous links. To put this general observation to the test, a potential solution to the current issues of the data market is developed in Chap. 4.

X. Oberson, A. E. Yazıcıoğlu, *Taxation of Big Data*,
https://doi.org/10.1007/978-3-031-43372-6_1

Once the scene is set, the tax law analysis begins. First, are examined the suggestions made to tackle the issues related to the taxation of data, within the context of the digital economy and on a stand-alone basis, in Chap. 5. Then, issues occurring in each *step* of the four-step methodology are examined step-by-step. *Step one* is analysed in Chap. 6, where the authors suggest implementing a brand new 'tax' to compensate users for the data they generate. *Step two* is examined in Chap. 7, where the authors succinctly describe the technological unemployment and make tax policy suggestions to alleviate its negative effects. *Step three* is delved into in Chap. 8, where the authors recommend implementing a new solution based on old tools. *Step four* is scrutinised in Chap. 9, where the authors explain their take on sale of data, transfer of data and the grant of the right to use a database. The tax law analysis, along with the book, ends with Chap. 10, concentrating on value added tax (VAT).

Preliminary Remarks This book opts for the use of (1) 'they' as a singular pronoun, in view of achieving sex neutrality; and (2) the term 'contributor' to designate all scientific commentators but the authors of this book. The term 'authors' is employed exclusively to refer to the authors of this book.

Chapter 2
Data and Their Life Cycle

Today's digital masters recognize that ultimately, "it's all about the data". [Malcolm et al. (2017), p. 60]

Abstract The raw material of the fourth industrial revolution is 'data'. When compared to the raw materials of previous industrial revolutions, this new raw material reveals to be more powerful and enhanced. In other words, 'data are not the new oil', they are 'oil on steroids'. This potency mainly derives from the five main characteristics of data: exponential value, low cost and effortless collection and distribution phases, non-rival nature, durable and reusable nature as well as non-excludable nature. Beginning by a single datum and turning into big data in the process, data have a life cycle that is typically composed of seven phases, namely generation, collection, storage, processing/analytics, distribution, use and deletion. Throughout this life cycle, data may take different forms, such as volunteered, observed, raw and processed. From datum to data, this new material constitutes an intangible. This chapter aims, thus, to introduce the youngest intangible forming the potent raw material of the fourth industrial revolution.

2.1 From 'Datum' to 'Big Data'

'Data' is the plural of 'datum', which refers to a piece of information.[1] A piece of information can be defined as an abstraction of a real-world entity (person, object or event).[2] The name of one specific person and the location information of a given person at a given time are examples of datum. 'Data', which is a collection of datum, can thus be defined as 'information'.

Different contributions analysing the impact of data tend to emphasise their digital nature. For example, the World Economic Forum summarises the concept

[1] Kelleher and Tierney (2018), p. 39.

[2] Kelleher and Tierney (2018), p. 39.

X. Oberson, A. E. Yazıcıoğlu, *Taxation of Big Data*,
https://doi.org/10.1007/978-3-031-43372-6_2

of personal data as "the digital record of 'everything a person makes and does online and in the world'".[3] In the same vein, the Organisation for Economic Co-operation and Development (OECD) explains the additional tidal wave of data by the digital recording of more and more offline activities.[4] Regardless of this tendency, to be qualified as data, 'information' does not necessarily have to be digital. The name, address and contact details of a person kept in a physical file also constitute 'data'. That being said, the data forming the raw material of the fourth industrial revolution,[5] or the second machine age,[6] are mainly digital. The data that give rise to tax law problems, and thus that constitute the subject of this book, are also predominantly digital.

The fourth industrial revolution is characterised by 'the digitisation of just about everything'.[7] Digitisation can be described as "the work of turning all kinds of information and media—text, sounds, photos, video, data from instruments and sensors, and so on—into the ones and zeroes that are the native language of computers and their kin".[8] 'The digitisation of just about everything', such as documents, news, music, photos, videos, personal updates and maps, is one of the most important phenomena of our era.[9] Data, the raw material of the latest industrial revolution, form one of the most significant outcomes of the digitisation.

Internet connectivity, the World Wide Web, search engines, smartphones, computers and applications (most commonly referred to as 'apps') enable the generation and the collection of substantial amounts of data. Less noticed but not less important in data generation are devices, other than smartphones and computers, that are connected to Internet, such as coffee machines, vacuum cleaners, cars and airplanes. The large-scale digitisation of daily and mundane devices was rendered possible by the developments that occurred in the sensor industry. When better performing yet cheaper and smaller sensors became available, it has become possible to embed them into almost every object.

Embedding the daily objects with sensors paved the way for a new type of connection, i.e., the device-to-device connection, which forms an additional layer to the existing person-to-person connection and person-to-information/services connection.[10] This connection between the devices is commonly referred to as 'the Internet of Things' (IoT). The phenomenon of IoT can be described as the

---

[3]World Economic Forum (2011), p. 13.

[4]OECD (2013), p. 8.

[5]Klaus Schwab describes the fourth industrial revolution as a period that is "characterised by a much more ubiquitous and mobile internet, by smaller and more powerful sensors that have become cheaper, and by artificial intelligence and machine learning.", Schwab (2016), p. 7.

[6]The term used by Erik Brynjolfsson and Andrew McAfee. See Brynjolfsson and McAfee (2014). This book opts for the term 'the fourth industrial revolution', since it is the most commonly used term, connecting more clearly with the historical approach.

[7]Title of the Chapter 4 of Brynjolfsson and McAfee (2014), pp. 57 ff.

[8]Brynjolfsson and McAfee (2014), p. 61.

[9]Brynjolfsson and McAfee (2014), p. 66.

[10]Meisel (2013), p. 10.

embedding of computing capability into devices and objects that have previously not had such capacity, and then the connecting of them to the Internet.[11] 'Ubiquitous computing' is an alternative term used for the same phenomenon. With this new era, humans began living in an environment of computation, as opposed to using computation, and digital services became integrated so seamlessly into their daily existence that no one even notices their presence.[12]

Digitisation made, thus, huge amounts of data readily accessible.[13] The use of the adjective 'huge' does not tantamount to an overstatement. The following example illustrates in a very clear manner how enormous is the amount of data that is being created on a daily basis. The complete works of William Shakespeare, in a basic text format stored on computer, would consume approximately five megabytes. A typical Airbus A350, on the other hand, generates approximately 2.5 terabytes of data per day, by means of roughly 6000 sensors embedded across the entire plane. Given that a gigabyte is 1024 megabytes and a terabyte is another 1024 gigabytes; the average A350 is producing the equivalent of 524,000 times the entire work of Shakespeare every 24 h.[14] Put in a larger perspective, as per the current estimates, 1145 trillion megabytes of data are created every day.[15] This number is expected to grow substantially by 2025. According to the estimates, 463 exabytes of data will be generated each day by people as of 2025.[16] "An exabyte is a ridiculously *big* number".[17]

It is therefore not surprising that the reference is frequently being made to 'big data'. While a precise definition of the term that has widely been agreed upon is currently lacking, the concept of 'big data' tends to be commonly used to describe large pools of data. Despite this common understanding, the volume is not the only defining feature of big data, which have three properties, frequently referred to as 'the three Vs',[18] namely volume, velocity and variety. Volume refers to ever-growing exabyte data rendered by machines, networks, and human interaction; velocity relates to the momentum of data creation, process, and storage; and variety describes the spectrum and intricacy of data varieties and data sources.[19] These properties are technical in nature, they depend on the evolution of data storage and processing technologies.[20] The more companies and organisations have become

---

[11] Malcolm et al. (2017), p. 22.

[12] Meisel (2013), pp. 10–11; Kerr and Szilagyi (2016), p. 349, footnote 93.

[13] Brynjolfsson and McAfee (2014), p. 67.

[14] Malcolm et al. (2017), p. 66.

[15] Bulao (2022).

[16] Bulao (2022).

[17] Brynjolfsson and McAfee (2014), p. 66. Emphasis made by the authors. 1 exabyte equals to 1024 petabytes and 1 petabyte equals to 1024 terabytes.

[18] Doug Laney, a Gartner analyst, was the first to coin the famous three Vs in 2001. Hammer et al. (2017), p. 8.

[19] Hammer et al. (2017), p. 8.

[20] OECD (2013), p. 12.

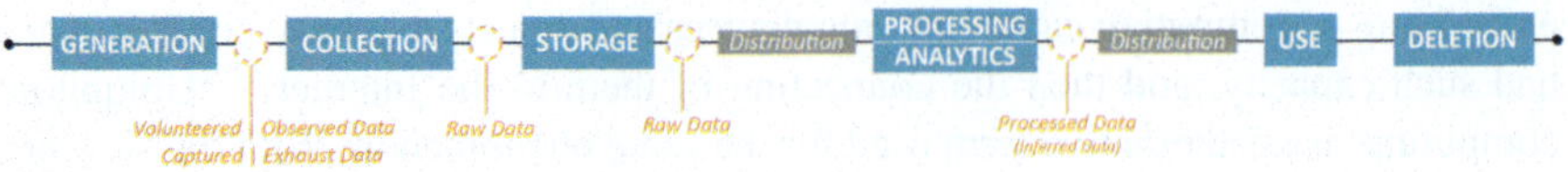

Fig. 2.1 Data's life cycle

acquainted with big data and incorporated them into their business operations; the more 'V's have come into existence. For instance; (1) veracity underlines the potential risk of bias and error in the data as one of the most significant challenges in obtaining value and validity; (2) volatility is associated with the ever-changing technology and business settings in which big data are created;[21] and (3) value refers to the increasing socioeconomic benefits that may derive from the employment of big data.[22]

For the sake of clarity, this book opts for the use of the concepts of 'data' and—to the extent relevant—'datum'. The main emphasis is on 'user data', i.e., data generated by or collected from natural persons, which form the backbone of the data business in its current state.

2.2 Life Cycle and Main Types of Data

The main phases of data's life cycle are as follows: (1) generation, (2) collection, (3) storage, (4) processing, (5) distribution, (6) analytics, (7) use and (8) deletion (see Fig. 2.1).[23] Expressing the life cycle of data by means of a linear process with clear-cut stages certainly amounts to an oversimplification. The most problematic aspect is the order of processing, distribution and analytics phases, which may be subject to change. In a similar vein, depending on the circumstances, some phases may not exist with regard to particular data. Types of data resulting from each phase may also vary. A given type of data may be created in one phase or in another phase; similarly, certain data can be repurposed and used for an entirely different objective. This section aims to describe the main phases of data's life cycle as well as the main types of data, as clearly as possible and to the extent relevant for the purposes of the tax law analysis conducted throughout this book. The explanations provided do not, and are not intended to, amount to technical explanations.

Data are usually generated by users, who are natural persons. As a matter of fact, the digitisation has given rise to a tremendous capacity to enable 'crowd sourcing' of

---

[21] Hammer et al. (2017), pp. 8–9.

[22] OECD (2013), p. 12.

[23] See also OECD (2013), p. 7. The 'use' phase is not mentioned by the OECD and the 'deletion' phase is not included in Fig. 1 of the document, entitled 'The data value chain and life cycle', due to the fact that, according to the OECD, the 'deletion' phase is more pertinent for protection of personal data and less important in the context of 'big data'.

user data.[24] The *generation and collection phases* are closely linked to each other. A generated datum is usually collected by one or several data aggregators in a simultaneous manner. Depending on the specificities of their generation and collection, data obtained during these two initial phases can be characterised as (1) either volunteered or observed; and (2) either captured or exhaust. For instance, a certain datum can be volunteered and captured.

The term '*volunteered data*' refers to information intentionally shared by natural persons especially through electronic media, such as social media (where users can share their birthday, occupation, marital status, photographs etc.) and e-commerce platforms (where users can share their address, credit card information, product preferences and opinion on purchased products).[25] By contrast '*observed data*' refer to the data gathered through the recording of users' activities, generally beyond their knowledge and without their intentional collaboration. Such data are mainly captured by means of consumers' behaviours and activities. They may be collected via a variety of technological tools, such as cookies, flash cookies, Global Positioning System (GPS) tracking and history sniffing.[26] Observed data are not limited to online activities. The devices equipped with sensors allow tracking the offline activities of users as well.

The term '*captured data*' refers to data intentionally collected by a data aggregator, i.e., data captured through a direct measurement process designed for data-gathering.[27] The term '*exhaust data*', on the other hand, is used to describe data that are a by-product of a process whose primary purpose is not capturing data. For example, for every image shared and tweeted, a range of exhaust data is generated, such as when it was shared, by whom it was shared and what device was used.[28]

While the data generated by users can amount to '*personal data*' in the legal sense of the term, this does not necessarily have to be the case. The term personal data is defined as "any information relating to an identified or identifiable natural person" in article 4(1) of the General Data Protection Regulation (GDPR).[29] As per the same paragraph, "an identifiable natural person is one who can be identified, directly or indirectly, in particular by reference to an identifier such as a name, an identification number, location data, an online identifier or to one or more factors specific to the physical, physiological, genetic, mental, economic, cultural or social identity of that natural person". It is not possible to provide an exhaustive list of personal data. Any datum that can identify a natural person can be classified as a personal datum. While

---

[24] OECD (2013), p. 7.

[25] World Economic Forum (2011), p. 14.

[26] Thimmesch (2016), p. 6.

[27] Kelleher and Tierney (2018), p. 240.

[28] Kelleher and Tierney (2018), p. 242.

[29] Regulation (EU) 2016/679 of the European Parliament and of the Council of 27 April 2016 on the protection of natural persons with regard to the processing of personal data and on the free movement of such data, and repealing Directive 95/46/EC. The reference is made to the GDPR, since the Regulation constitutes the most comprehensive personal data protection regulation to the extent known by the authors.

names, credit card numbers, addresses and photographs are among the most obvious examples; Internet Protocol (IP) addresses,[30] nicknames, handwritings and records of working time[31] may also be considered as personal data. As evidenced by its definition, 'personal data' has, in fact, been designed as a dynamic concept. The emergence of new technologies and types of data will inevitably have the effect of enlarging its scope. The circumstances of a particular case may also be pertinent in determining whether a datum should be considered a personal datum.

The main purpose of personal data regulations is to protect the privacy of natural persons. The issues examined in such regulations are, overall, not pertinent for the purposes of the analysis conducted in this book. For this reason, throughout this manuscript, personal data are not distinguished from other data, unless such a distinction reveals to be necessary for the purposes of the legal analysis.

Data generated and collected during the first two initial phases of data's life cycle constitute '*unstructured data*', that are commonly referred to as '*raw data*'. The concept of 'raw data' is fairly self-speaking; it simply points to the fact that data have been piled up, without any processing or refinement. Raw data can thus be defined as pile of actual data elements.[32] While the term 'unstructured data' is more appropriate for academic purposes in the authors' opinion (especially given the fact that after processing and analytics phases, data are classified as 'semi-structured' and 'structured'), the term 'raw data' is employed throughout this contribution for two reasons. First, the terms 'semi-structured data' and 'structured data' are regrouped under the term 'processed data', as it is explained below. Second, most contributions opt for the term 'raw data'.

The raw data that have been collected by a data aggregator have to be *stored* until they are processed, distributed or even destroyed without being used. With regard to data, especially before the entry into force of personal data regulations, companies seem to have followed a 'save-it-all' mentality for a non-negligible time frame. Such an approach was supported by the perceived low cost of storing data. As a matter of fact, storage costs dropped significantly in the last two decades and storage capacity increased in a tremendous manner. These two trends have given rise to the impression that data storage is extremely cheap. This impression is, however, not accurate. There are many hidden costs of data storage, to name a few: expenses related to power, cooling, noise reduction, security, backup management and proper disposal. In countries where robust personal data regulations are in force, data collection, and thereby storage, has to be reduced. Other legislation regulating how organisations must deal with particular types of data, such as legislations related to compliance and

[30] One of the Court of Justice of the European Union (CJEU)'s landmark judgments in this regard is Case C-70/10 [2011] *Scarlet Extended SA v Société belge des auteurs, compositeurs et éditeurs SCRL (SABAM)*, ECLI:EU:C:2011:771.

[31] See the judgment of CJEU, Case C-342/12 [2013] *Worten — Equipamentos para o Lar SA v Autoridade para as Condições de Trabalho (ACT)*, ECLI:EU:C:2013:355.

[32] Duch-Brown et al. (2017), p. 38.

insurance, may also produce a similar decrease. That being said, data aggregators still seem to collect and store substantial amounts of unnecessary data.[33]

The 'unnecessary data' are commonly referred to as '*ROT data*', i.e., 'Redundant, Obsolete and Trivial' data. ROT data do not have any value. According to the Databerg Report, prepared by Veritas Technologies LLC[34] in 2016, the global average of existent ROT data was 33%.[35] While ROT data were estimated to form 33% of the existing stored data, the real percentage was even higher. As a matter of fact, a part of the stored data is formed by '*dark data*', i.e., unknown data, which may consist of pertinent data that can be used and/or ROT data.[36] As per the Databerg Report, the global average of stored dark data amounted to 52% in 2016.[37]

The collected data do not necessarily have to be stored by the data aggregator itself. A data aggregator can have recourse to storage services provided by another entity. Employment of such services, which are available both for analogue and digital data, are quite frequent in practice. One of the most commonly used storage services for digital data is cloud storage.

An entity that has collected and stored raw data may either sell or—in the context of intragroup companies—transfer them to another entity (the *distribution* phase) or process them to be able to use them. During the *processing and analytics* stages, data are being 'refined'. As a matter of fact, data analysis can be defined as "any process for extracting useful information from data".[38] During these stages, may be produced, among others, data visualisation, summary statistics, correlation analysis and data sets. A data set can be defined as "a collection of data relating to a set of instances, with each instance described in terms of a set of attributes".[39] The term 'attribute' is often used interchangeably with the terms 'variable' and 'feature' to denote an individual abstraction.[40] For instance, a book has the following attributes: author(s), title, publisher, price, publication date, etc.[41]

During the processing and analysis phases, '*inferred data*' can be created. Inferred data are data about users that are based on the analysis of the volunteered and/or observed data obtained from/about the users concerned.[42] For example, on the basis of information relating to an individual's financial history (income, shopping patterns, mortgage information, etc.), credit scores can be calculated.[43] In the

[33] Sergeant and Sergeant (2010), pp. 41–47; Dutta and Hasan (2013), pp. 29–43.

[34] An American international data management company founded in 1983.

[35] Veritas Technologies LLC (2016), p. 4.

[36] Veritas Technologies LLC (2016), p. 3.

[37] Veritas Technologies LLC (2016), p. 4.

[38] Kelleher and Tierney (2018), p. 240.

[39] Kelleher and Tierney (2018), p. 241.

[40] Kelleher and Tierney (2018), p. 39. The term 'attribute' is often used interchangeably with the terms 'variable' and 'feature' to denote an individual abstraction.

[41] Kelleher and Tierney (2018), p. 39.

[42] World Economic Forum (2011), p. 7.

[43] World Economic Forum (2011), p. 14.

same vein, on the basis of an individual's recent web searches and purchases, it can be deduced that the individual concerned is suffering from depression. Inferred data, thus, constitute a type of educated guess about an individual.[44] Needless to state, depending on the entity using the data and the circumstances of the particular case, inferred data may also be created during the *use* phase.

At the end of the processing and analytics stages, raw data are 'refined' into *semi-structured* and *structured* data. Semi-structured data contain some structural elements (like tags) unlike unstructured (raw) data, but do not follow the format of a tabular data model or relational databases.[45,46] Structured data, on the other hand, refer to data with strong and consisted organisation,[47] which means data that can be stored in a table.[48] These two different types of data are regrouped under the term '*processed data*' for the purposes of this contribution. As a matter of fact, the semi-structured or structured nature of data does not affect the tax law analysis conducted in this book. The term 'processed data' is also frequently used by most contributions written in the legal field. In fact, employment of the terms 'semi-structured' and 'structured' data seem to be rare when compared to 'processed data'.

An entity that has created processed data may either sell or—in the context of intragroup companies—transfer them to another entity (the *distribution* phase) or to use them for its own internal, typically research and development (R&D) and commercial, activities (the *use* phase). Potential use cases of processed data are described, to the extent relevant, in Chap. 3. After their initial distribution or use, data can be processed once more and be used for an entirely different purpose. For example, a company may proceed to the anonymisation of personal data that were initially used for commercial purposes and utilise them in the framework of statistics, or sell them to another company.

The life cycle of data concludes with *deletion*. Data that are no longer needed can be deleted from the system of the data aggregator. In the absence of a legislation requiring the deletion of a certain category of data, nothing prevents the data aggregator from keeping the data, as long as it deems appropriate to do so, for potential ulterior use.

---

[44] Thimmesch (2016), p. 7.

[45] Database can be defined as central repository of data. 'Relational database', which stores data in tables with a structure of one row per instance and one column per attribute, is the most common database structure. Such a representation is, in fact, ideal for storing data with a clear structure that can be decomposed in national attributes. Kelleher and Tierney (2018), pp. 240–241.

[46] Teradata (2021a).

[47] Teradata (2021b).

[48] Kelleher and Tierney (2018), p. 245.

2.3 Data Are 'the Oil on Steroids'

Data are being frequently compared to 'oil' or 'fuel'. Despite some—unnecessarily—strong reactions,[49] which were generally published online, this comparison is clearly not used with the intention to make a technical statement. No expert opinion is needed to understand that the properties of data and oil diverge to a great extent. By having recourse to this analogy, the contributors in the field are merely aiming to underline the two common aspects of data and oil (developed below), which are not of a negligible nature. This is one of the reasons for which the analogy is examined in this book. The second reason for incorporating this brief analysis is to demonstrate the main characteristics of data, which diverge to a great extent from other assets and which can be effortlessly understood when explained in comparison to a material good, like oil.

The phrase 'data is the new oil' seems to have been coined almost two decades ago (in 2006) by Clive Humby, a British mathematician and the architect of The Tesco Clubcard, a supermarket reward program. Humby's analogy was based on the fact that data need 'refinement' to be used, just like crude oil.[50] The data-oil analogy seems to have become extremely popular when an article entitled 'The world's most valuable resource is no longer oil, but data' was published in the Economist in 2017.[51] The overuse of the analogy resulted in an ever-increasing expression of counter-arguments and 'more appropriate' analogies. It has been suggested, among others, that data are not the new oil but 'the new nuclear power'[52] and that data are more akin to renewable sources of energy, such as sunlight, wind and tides.[53]

The data-oil analogy successfully underlines the two common aspects of these materials. First, data are the primary raw material of the fourth industrial revolution, just as coal was the primary raw material of the first industrial revolution; and oil, steel and electricity were the 'new' raw materials of the second industrial revolution.[54] Second, just like oil, data need to be 'mined', 'refined' and 'distributed'.[55] These similarities are not inconsequential, not only from a general perspective but also from a tax law perspective.

By means of the data-oil analogy, it is possible to explain and illustrate the five main characteristics of data, namely: (1) exponential value; (2) low cost and effortless collection and distribution phases; (3) non-rival nature; (4) available amount; and (5) non-excludable nature.

---

[49] Such reactions, or their references, are not reproduced in this book.

[50] Bridle (2018).

[51] The Economist (2017).

[52] Bridle (2018).

[53] Marr (2018).

[54] Malcolm et al. (2017), p. 27.

[55] Malcolm et al. (2017), p. 65.

Exponential Value The value of one barrel of oil, which is a fungible commodity, is roughly equivalent to the value of another barrel of oil. In other words, oil's value is linear. Once the value of one barrel of oil is known, the value of 10 or 100 barrels of oil can be calculated without any difficulties, by simply multiplying the value of one barrel with the number of barrels.[56] The same does not hold true for data, which are non-fungible.[57] Data's value, which may be unknown at the time of collection, depends on several factors, such as their particular properties, volume, method of use, commercialisation potential and combination with other data sets.[58] Data's value is thus exponential. The value of 100 terabytes of data cannot be calculated simply by multiplying the value of 1 terabyte of data with 100. As a matter of fact, while 1 terabyte of data is worth $1\times$, 10 terabytes may be worth $10\times^2$ and the value of 100 terabytes may amount to $100\times^n$.[59]

Low Cost and Effortless Collection and Distribution Phases Mining oil is a localised and expensive activity, whereas creation and capture of data is distributed and becomes much cheaper after the initial investment has been made. In the same vein, while distribution of oil requires pipes, ships and similar equipment, distribution of data occurs via cables, fiber-optic networks or similar equipment. Producing and distributing an additional barrel of oil generates a significant expense, while generating, storing and transferring data becomes nearly costless once an initial expense to set up the relevant system has been made.[60] In more technical terms, oil has positive marginal costs; whereas data have zero marginal cost.

Non-rival Nature Material goods—like oil—are by their very nature rival goods, which means they can only be used by one person at a time.[61] For example, the hard copy of a book can be read only by one person at a time. The book is, of course, not 'used up' and can be used again after this initial use; however, it cannot be simultaneously used by two different persons. Some rival goods can only be consumed once, such as oil. Once a barrel of oil is consumed, it cannot be consumed again. On the other hand, immaterial goods and assets, such as data, are non-rival. A number of people can use the same data at the same time without any loss of content for any of the parties involved.[62] Also, data do not perish with 'consumption'. Quite the opposite, not only they can be used over and over again for the same or different purposes, but also, the more they are utilised the more they become useful. As a matter of fact, once they are processed, further possible applications of data can be revealed.[63]

---

[56] Malcolm et al. (2017), p. 67.

[57] Stapp (2019).

[58] Stapp (2019).

[59] Malcolm et al. (2017), p. 67.

[60] Malcolm et al. (2017), p. 67; Stapp (2019).

[61] Duch-Brown et al. (2017), p. 12.

[62] Duch-Brown et al. (2017), p. 12. See also, Janeček (2018), p. 1045; and OECD (2019), p. 17.

[63] Marr (2018).

Available Amount Oil is finite; there exists a certain amount of oil, which will not increase with time. Data, on the other hand, are infinitely durable and reusable.[64] They can be generated continuously, as needed. In this respect, data are, indeed, more akin to renewable sources, like sunlight and wind.[65]

Non-excludable Nature Oil is excludable, whereas data are not. Oil can be stored in a manner that will prevent its use by non-authorised parties. On the other hand, generally speaking, it is not possible to exclude the collection and the use of similar, or the same, data by third parties.[66]

The cumulation of non-rivalry and lack of excludability tends to indicate the existence of a public (common) good.[67] By its very nature, a public good can be used by everyone and a person's employment of a public good does not reduce the ability of other persons to make use of it. For instance, a road can be used by everyone and the fact that a number of persons are using the road will not, *per se*, reduce the ability of other persons to use it. Some scholars already underlined that data form a common good.[68] While this observation made on the basis of the intrinsic characteristics of data may hold true for certain types of data (such as news and meteorological information);[69] due to the current functioning of the data market, as well as certain existent legislations in some countries, data may often act as a private good, which is both rival and excludable.

As a matter of fact, and as analysed in detail in Chap. 3, a non-negligible portion of data is collected by private companies and is simply not available to the general public. This *modus operandi* renders the data market nothing but rivalrous, especially with regard to data relating to users. Given the 'battle of digital giants'[70] for a quasi-exclusive collection of user data, access to such data may even reveal to be precluded for potential competitors of the digital giants trying to enter the relevant market.[71] Data can, moreover, be rendered excludable either by means of legislation or through measures taken by private parties. For example, the data privacy

[64] Marr (2018).

[65] Marr (2018); Malcolm et al. (2017), p. 67.

[66] Stapp (2019); See also, Franceschi and Lehmann (2015), pp. 54–55. It is interesting to note that some scholars have put forward proposals to construct an exclusive data control and grant data a rival character. 'User held data ownership' (i.e., data collected in a 'personal data cloud') and 'Databox' proposals exemplify this point of view. For further information see, Jurčys et al. (2020) and Haddadi et al. (2015).

[67] The term 'public good' is not used to indicate that data form goods and not services from a tax law point of view. The reference is being made to the economic term of 'public good'.

[68] On this point, see, for example, Swinnen (2020), p. 160; Hummel et al. (2021), p. 549; Singh and Vipra (2019), p. 56; and Prainsack (2019), p. 4.

[69] News and meterological information are used for illustrative purposes only. Depending on the circumstances, such data may also form 'private goods'.

[70] 'The digital giants' and other similar terms traditionally refer to Amazon.com, Apple, Facebook, Google and Microsoft.

[71] Purtova (2015), pp. 107–109.

regulations can be interpreted as amounting to grant data subjects a level of exclusiveness relating to their personal data, especially through the right to require the deletion of personal data.[72] In a similar vein, certain data may be rendered excludable by employment of encryption tools or other high-degree protection mechanisms.[73]

It is undisputed that rivalry and excludability do not constitute intrinsic characteristics of data *per se*. The existent 'deviations' are exclusively linked to the functioning of the market or the applicable legislations and do not alter the main features of the concept of data. That being said, under the current scheme of things, some data are clearly acting as a private good. Such data form the subject matter of this book.

On the basis of this brief analysis, it is possible to conclude that data are 'the new oil' in certain respects, while being—overall—superior to oil as a raw material. The authors, therefore, argue that data are 'the oil on steroids'.

2.4 Data: The Youngest Intangible

2.4.1 Legal Framework Applicable to Data

To be able to qualify data from a tax law perspective, first the general legal framework relating to data needs to be examined. The legal framework relating to collection of, use of, and transactions in data remains highly undeveloped to date.[74] Determining legal ownership rights in data (and identifying to what extent such rights should be established), offering the most appropriate legal framework to protect personal data and deciding whether data could and/or should be protected by an intellectual property right (and in the affirmative what kind of right, which types of data and to what extent) continue to be heavily debated.[75] Most jurisdictions do not have any specific legal framework applicable to data, other than in the area of regulating data privacy.[76]

The general tendencies with regard to legal ownership rights in data and intellectual property frameworks were summarised by Sprague in the General Report prepared on the subject of "Big data and tax- domestic and international taxation of data driven business" for the International Fiscal Association's 2022 Congress. Given the fact that the General Report is based on the national reports prepared by

---

[72] Victor (2013), p. 524; Boerding et al. (2018), p. 331. On this point, see also Janeček (2018), p. 1045.

[73] Kerber (2016), p. 761. In the same vein, see Duch-Brown et al. (2017), p. 15 where the contributors state that "secrecy and technical access restrictions can ensure data excludability."

[74] Sprague (2022), p. 18.

[75] For a brief overview of the EU perspective, see Duch-Brown et al. (2017), pp. 12–19.

[76] Sprague (2022), p. 18.

domestic experts in tax law and it was published in June 2022, it has the merit of reflecting a rather up-to-date[77] overview of the legal framework on a global scale. The findings of the Report can be summarised as follows:[78]

- There is no property right in data *per se*, even when data have been aggregated in a database. Various property law concepts may, however, apply to protect against access to data and information by third parties.
- While no copyright protection is offered to data, some jurisdictions (such as EU and the US) offer copyright protection to databases to a certain extent.
- Some jurisdictions provide for limited *sui generis* protection for collections of valuable data sets.
- Device manufacturers can generally not claim trade secret ownership rights in data collected by the devices they sell to consumers. Consumers typically cannot claim trade secret rights in the data collected by the devices they own either, due to the fact that they cannot normally claim a competitive advantage from keeping such data secret.
- While parties are free to regulate the use of data provided under contract per the terms of the agreement, contract law does not convey actual property rights.
- Government regulation can impose obligations on those who acquire and process data, which is typically the case for personal data. Such obligations may be interpreted as amounting to a residual right in that data, which is a property right. For example, by granting specific rights to individuals on their personal data, the GDPR has the effect of attributing any remaining residual rights to the data collector, however limited they may be due to the collector's fiduciary role.[79]
- Unique control over specific data sets may allow some enterprises to commercialise the data concerned from a practical standpoint. Control may be exercised, to a certain extent, by not allowing others to access the data or by transferring the data only under limitations provided by contract. Must be borne in mind, however, that a contract is only enforceable between the parties and not towards third parties who are not signatories. This implies that once the data are out in the open, the data 'owner' has no legal means to enforce its rights.[80]

Discussing the potential solutions to tackle the 'uncertainty' reigning in the legal ownership rights in data (to the extent where such 'uncertainty' can and is intended to be resolved[81]), to implement the most efficient personal data protection framework or to delineate the specificities of an appropriate intellectual property right

[77] This manuscript was finalised in July 2023.

[78] The following information recapitulated in bullet points was mainly taken from Sprague (2022). Two additional pieces of information were added to the conclusions of the General Report. The source of this additional information is indicated in footnotes.

[79] This aspect is not mentioned in the General Report. See Duch-Brown et al. (2017), p. 17.

[80] Duch-Brown et al. (2017), p. 15.

[81] Countries that are home to digital giants may typically be reluctant to acknowledge the existence, or the conceptualisation, of any such category of data ownership rights. On this point, see Singh and Vipra (2019), p. 55.

regime (again, to the extent that it is needed) is beyond the scope of this publication.[82] Such issues are certainly relevant for tax purposes and produce a non-negligible impact on the tax treatment of sale of data, transfer of data and the grant of the right to use a database.[83] Nevertheless, they do not otherwise radically modify the potential tax law solutions that are suggested in this book to remedy to the existing problems of taxing data. In fact, regardless of the legal uncertainty over their ownership, data are indubitably subject to *de facto* ownership.[84] This latter form of ownership constitutes a viable basis from a tax law perspective, due to the well-established substance-over form approach.[85] Any legal clarification with regards to ownership and protection of data will have the merit of clarifying the tax treatment of data, and related transactions, within the designated fiscal framework. The purpose of this book is to develop an easy-to-implement and efficient tax policy and framework, which can be put in place in a rather swift manner to remedy to the shortcomings of the current system. Like every new proposal, the suggestions made by the authors can be developed and modified to a certain extent in the future, in accordance with technological and legal developments.

2.4.2 *Qualification of Data from a Tax Law Perspective*

As per the International Accounting Standard (IAS) 38 (entitled 'intangible assets'), "an intangible asset is an identifiable non-monetary asset without physical substance".[86] The Standard mentions as examples of intangibles, among others, computer software, customer lists, customer or supplier relationships, market share and marketing rights.[87] While the terms 'non-monetary asset' and 'asset without physical substance' can be understood without major difficulties, the 'identifiability' of assets necessitates further explanations. "An asset is identifiable if it either: (a) is separable, ie is capable of being separated or divided from the entity and sold, transferred, licensed, rented or exchanged, either individually or together with a related contract, identifiable asset or liability, regardless of whether the entity intends to do so; or (b) arises from contractual or other legal rights, regardless of whether those rights are transferable or separable from the entity or from other rights and obligations".[88]

[82] For the discussion on the intellectual property protection of data see for example, Kerber (2016), p. 761; Swinnen (2020), p. 165; Duch-Brown et al. (2017), p. 14; Heverly (2003), pp. 1130–1333; and Fadler and Legner (2022), p. 124.

[83] Examined in Chap. 9.

[84] See for example, Duch-Brown et al. (2017), p. 17; Swinnen (2020), p. 168; and Boerding et al. (2018), p. 352.

[85] Vanistendael (1996), pp. 41–60.

[86] IAS 38, § 8.

[87] IAS 38, § 9.

[88] IAS 38, § 12.

In its Transfer Pricing (TP) Guidelines, the OECD defines intangibles as "something which is not a physical asset or a financial asset, which is capable of being owned or controlled for use in commercial activities, and whose use or transfer would be compensated [. . .]".[89]

It can, therefore, be stated that to be qualified as an intangible asset, data should be: (1) a non-monetary asset; (2) without physical substance; (3) capable of being separated or divided from the entity and used in commercial activities, regardless of whether the entity intends to do so; and (4) subject to compensation when used or transferred.

2.4.2.1 A Non-monetary Asset

A monetary asset is defined as "money held and assets to be received in fixed or determinable amounts of money" in the IAS 38.[90] *Data cannot be qualified* as money held or assets to be received in fixed or determinable amounts of money.

It can be observed that data have been referred to as 'a currency', mainly due to the fact that they can be used in lieu of money to receive digital services and that the data-driven economy continues to grow.[91] Referring to data as 'a currency' allows to emphasise the massive amounts of data that are being collected from the users and their substantial value, even if this was not necessarily the intention of the contributors who made recourse to such a reference.

As a matter of fact, data cannot fulfill the three cumulative functions of money, namely being used as a medium of exchange, unit of account and store of value.[92] Money is the commonly accepted medium of exchange, which means that a person can purchase goods and services available on the market, such as a book and an airplane ticket, with money. Clearly, in principle, data cannot act as such a medium of exchange. While data can be exchanged for a number of digital services,[93] to date, it is not possible to purchase a book or an airplane ticket, or the majority of goods and services available on the market, in exchange for data. A unit of account can be defined as measurement unit of the market value of goods and services. For instance, an airplane ticket may cost USD 250 and a book may cost USD 25. It would be rather impossible to express the value of goods and services in terms of data, at least currently. Last but not least, money acts as a store of value, it can be spent or it can be

[89] OECD TP Guidelines, § 6.6.

[90] IAS 38, § 8.

[91] See, for example, Ben-Shahar (2019), p. 142; Marian (2022), pp. 32–34; and World Economic Forum (2011), p. 13.

[92] Mankiw (2017), p. 605. Some key economic terms and principles are employed throughout this book. Definitions of such terms and principles as well as explanations relating to them can be found in a variety of contributions on economics. The reference is made to one contribution per concept throughout the book.

[93] As a matter of fact, data may be regarded as a consideration for 'free' digital services in the context of the so-called 'digital barter'. For more details, see Chap. 3, Sect. 3.5.1.

saved to be spent at a later stage, which would have the effect of transferring the purchasing power from the present to the future. As developed in Chap. 3, determining the value of data is not an easy exercise. Even when calculated in a rather accurate manner at a given time, the value of data is prone to fluctuate with time. It is, therefore, impossible to use data as a store of value.

Data thus carry the first feature of intangible assets.

2.4.2.2 An Asset Without Physical Substance

As expressed in Sect. 2.1 of this chapter, data are information. Information, by its very definition, does not have any physical substance. Data can, however, be contained in or on a physical substance, such as a compact disc, a computer or paper. To the extent where data do not form an integral part of the related physical element, which will generally be the case since the purpose of the physical element is merely storing and/or transferring data, the physical substance of such an element is not relevant.[94] It can therefore be stated that the immaterial character of data persists even when they are contained in a data carrier that has a physical substance. Consequently, data carry the second feature of intangible assets as well.

From an accounting perspective, it is also important to determine whether the data carrier may lose its material character due to the data it contains. This question is to be replied by the affirmative if the data career is of subordinate importance to the data it holds. This will especially be the case if the functions of the data carrier are limited to storing data, such as a USB stick.[95] In order to be able to assume subordinate importance, it is not only necessary to focus on the dependency of the functionality, but also to consider the other functions that belong to the data carrier in question. For example, a computer whose hard disk can also be used as a data storage medium cannot be subordinated to the data—and treated as an intangible asset—on this basis, since the device concerned is not used for the sole purpose of storing data.[96]

2.4.2.3 An Asset Capable of Being Separated or Divided from the Entity

Data are capable of being separated or divided from the data aggregating entity and sold, transferred or exchanged. A non-negligible number of data aggregating entities may not intend to sell or transfer their data, but this lack of intention does not affect the intangible character of data as such.[97]

It is beyond any doubt that a data aggregator can transfer or sell a part of its raw data, a part of its processed data and even a single datum that it has. As a matter of

[94] IAS 38, § 4–5.

[95] Schwarz (2019), pp. 132–133.

[96] Schwarz (2019), p. 132.

[97] IAS 38, § 12.

fact, nothing excludes a single datum from being traced, identified, separated and transferred or sold. The fact that such a separation does not generally occur should not be interpreted as producing an impact on the intangible character of datum.

Thereupon, data also carry the third feature of intangible assets.

2.4.2.4 An Asset Subject to Compensation

Beyond any doubt, a single datum, raw data and processed data all have value and all should normally be subject to compensation. The fact that some types of data are currently not subject to compensation is not relevant and does not modify the intangible nature of datum and data. Since the value of data is discussed in detail in Chap. 3, no further explanations are provided here.

In this regard, "Transfer Pricing" Working Group argued that a single item of data (i.e., datum) would normally not be considered as an intangible due to the absence of a potential resource inflow. The Working Group underlines that while a datum contributes to the value of the entire data set, it is likely to be worthless when assessed on its own.[98] As examined in Chap. 3, the authors are of the opinion that from datum to big data, data have value. Beyond any doubt, the value of a datum is substantially low when compared to the value of a data set. As a matter of fact, the value of a datum would be so inferior that it should not even be compared to a data set to begin with. Nonetheless, a datum has some value, even if it is for the sheer reason that a data set cannot exist without a group of datum. Also, denying the intangible character of datum, while arguing that a collection of datum needs to be qualified as an intangible, does not seem to constitute the most suitable approach.

Data, thus, carry the fourth feature of intangible assets as well.

2.4.2.5 Conclusion: The Youngest Intangible

Data, the 'new' raw material of the latest industrial revolution, carry all the features of intangible assets. This cumbersome to appraise, difficult to analyse from a legal standpoint, superior than oil source of 'superpower' forms, thus, the youngest (and maybe—so far—the coolest) intangible.[99]

Nota Bene This book mainly concentrates on user data, i.e., data generated by or collected from natural persons, which form the backbone of the data business in its

[98] Schmalenbach-Gesellschaft "Transfer Pricing" Working Group (2020), p. 492.

[99] As may be known by the readers, the conclusion reached on the basis of the analysis conducted in this section is not new and unique to this book. The same observation has already been made before by a number of scientific contributors as well as the OECD, which exemplifies 'customer data' as one of the marketing intangibles in its TP Guidelines [OECD TP Guidelines, § 6.6]. For more information on this point, see, for example; Boerding et al. (2018), p. 346; Kerber (2016), p. 760; and OECD (2019), p. 17.

current state. Such data do not necessarily amount to 'personal data' in the legal sense of the term. The primary purpose is to develop an efficient tax policy and framework for the data currently acting as a private good. Other legal issues, such as legal ownership, protection of personal data and intellectual property right regimes, are beyond the scope of this book and only mentioned to the extent pertinent for the tax law analysis.

References

Ben-Shahar O (2019) Data pollution. J Leg Anal 11:104–159

Boerding A, Culik N, Doepke C et al (2018) Data ownership a property rights approach from European perspective. J Civil Law Stud 11(2):323–370

Bridle J (2018) Opinion: data isn't the new oil-it's the new nuclear power. https://ideas.ted.com/opinion-data-isnt-the-new-oil-its-the-new-nuclear-power/. Accessed 07 July 2023

Brynjolfsson E, McAfee A (2014) The second machine age, 1st edn. W. W. Norton & Company

Bulao J (2022) How much data is created every day in 2022?. TechJury. https://techjury.net/blog/how-much-data-is-created-every-day/. Accessed 11 Aug 2022

Duch-Brown N, Martens B, Mueller-Langer F (2017) The economics of ownership, access and trade in digital data. JRC digital economy working paper 2017-01, European Commission Joint Research Centre, European Union

Dutta AK, Hasan R (2013) How much does storage really cost? Towards a full cost accounting model for data storage. In: Altmann J, Vanmechelen K, Rana OF (eds) Economics of grids, clouds, systems, and services. GECON 2013, Lecture Notes in Computer Science, vol 8193. Springer, Cham, pp 29–43

Fadler M, Legner C (2022) Data ownership revisited: clarifying data accountabilities in times of big data and analytics. J Bus Anal 5(1):123–139

Franceschi AD, Lehmann M (2015) Data as tradeable commodity and new measures for their protection. Italian Law J 1(1):51–72

Haddadi H, Howard H, Chaudhry A (2015) Personal data: thinking inside the box, ArXiv. https://arxiv.org/pdf/1501.04737. Accessed 07 July 2023

Hammer C, Kostroch DC, Quiros G (2017) Big data: potential, challenges and statistical implications. IMF staff discussion note, IMF

Heverly RA (2003) The information semicommons. Berkeley Technol Law J 18(4):1127–1189

Hummel P, Braun M, Dabrock P (2021) Own data? Ethical reflections on data ownership. Philos Technol 34:545–572

Janeček V (2018) Ownership of personal data in the Internet of Things. Comput Law Secur Rev 34:1039–1052

Jurčys P, Donewald C, Fenwick M et al (2020) Ownership of user-held data: why property law is the right approach. Harvard Journal of Law and Technology Digest. https://jolt.law.harvard.edu/assets/digestImages/Paulius-Jurcys-Feb-19-article-PJ.pdf. Accessed 07 July 2023

Kelleher JD, Tierney B (2018) Data science, 1st edn. The MIT Press Essential Knowledge Series

Kerber W (2016) Governance of data: exclusive property vs. access. Int Rev Intellect Prop Competition Law 47:759–762

Kerr I, Szilagyi K (2016) Asleep at the switch? How killer robots become a force multiplier of military necessity. In: Calo R, Froomkin MA, Kerr I (eds) Robot law, 1st edn. Edward Elgar, pp 333–366

Malcolm F, Roehrig P, Pring B (2017) What to do when machines do everything: how to get ahead in a world of AI, algorithms, bots, and big data, 1st edn. Wiley

Mankiw NG (2017) Principles of economics, 8th edn. Cengage Learning

Marian O (2022) Taxing data. Brigham Young Univ Law Rev 47(2):511–576
Marr B (2018) Here's why data is not the new oil. Forbes. https://www.forbes.com/sites/bernardmarr/2018/03/05/heres-why-data-is-not-the-new-oil/. Accessed 07 July 2023
Meisel W (2013) The software society: cultural and economic impact, 1st edn. Trafford Publishing
OECD (2013) Exploring data-driven innovation as a new source of growth: mapping the policy issues raised by "Big Data". OECD digital economy papers no. 222. OECD Publishing
OECD (2019) Enhancing access to and sharing of data: reconciling risks and benefits for data re-use across society. OECD Publishing
Prainsack B (2019) Logged out: ownership, exclusion and public value in the digital data and information commons. Big Data Soc 6(1):1–15
Purtova N (2015) Illusion of personal data as no one's property. Law Innov Technol 7(1):83–111
Schmalenbach-Gesellschaft "Transfer Pricing" Working Group (2020) Data and information as taxable assets. Eur Tax IBFD 60(11):489–498
Schwab K (2016) The fourth industrial revolution, 1st edn. World Economic Forum
Schwarz AM (2019) Die Handels- und Steuerrechtliche Behandlung von Daten, 1st edn. Stämpfli Verlag
Sergeant AMA, Sergeant CS (2010) Hidden costs of data storage. J Corp Acc Financ 21(5):41–47
Singh PJ, Vipra J (2019) Economic rights over data: a framework for community data ownership. Development 62:53–57
Sprague GD (2022) General report. In: International Fiscal Association 2022 Berlin Congress, Cahiers de droit fiscal international: big data and tax-domestic and international taxation of data driven business, vol 106B. International Fiscal Association, pp 11–71
Stapp A (2019) Why data is not the new oil. https://truthonthemarket.com/2019/10/08/why-data-is-not-the-new-oil/. Accessed 07 July 2023
Swinnen KK (2020) Ownership of data: four recommendations for future research. J Law Prop Soc 5:139–175
Teradata (2021a) What is semi-structured data? https://www.teradata.com/Glossary/What-is-Semi-Structured-Data. Accessed 07 July 2023
Teradata (2021b) What is structured data? https://www.teradata.com/Glossary/What-is-Structured-Data. Accessed 07 July 2023
The Economist (2017) The world's most valuable resource is no longer oil, but data. The Economist. https://www.economist.com/leaders/2017/05/06/the-worlds-most-valuable-resource-is-no-longer-oil-but-data. Accessed 07 July 2023
Thimmesch A (2016) Transacting in data: tax, privacy and the new economy. Denver Law Rev 94(1):145–194
Vanistendael F (1996) Legal framework for taxation. In: Thuronyi V (ed) Tax law design and drafting, vol I, 1st edn. IMF, pp 15–70
Veritas Technologies LLC (2016) The Databerg Report: see what others don't identify the value, risk and cost of your data. https://www.veritas.com/content/dam/Veritas/docs/reports/scd_veritas_strike_summary_a4-ls-usa_final.pdf. Accessed 07 July 2023
Victor JM (2013) The EU General Data Protection Regulation: toward a property regime for protecting data privacy. Yale Law J 123(2):513–528
World Economic Forum (2011) Personal data: the emergence of a new asset class. World Economic Forum

Chapter 3
Anatomy of the 'Data Business'

Abstract This chapter sketches the anatomy of the 'data business'. To this end, first, the five main actors of the data market, namely two or multi-sided platforms, data brokers, other data aggregators, users and microworkers are examined one by one. On the basis of this examination, it is concluded that while two or multi-sided platforms, data brokers and other data aggregators may generate substantial revenues from the use of data; users and microworkers constitute the hidden locomotives of the 'data business'. The observation of the interactions between the actors reveals the four main dynamics of the data market, which are, users' expectation of the digital services to be free; the fact that users are not always getting 'free' services in exchange for their data; domination of the market by a small number of companies; and the necessity of ameliorating the working conditions of microworkers. These dynamics translate into four main issues in the functioning of the data market, namely the misconception of raw data being 'valueless', privacy concerns, technical unemployment and disconnection of the place of business from the place of consumption. This chapter resumes by elucidating the 'four-step methodology' forming the backbone of the tax law analysis conducted throughout the book. Step one consists of data generation by users and data collection by data aggregators. Step two involves activities undertaken by microworkers. Step three comprises aggregation and internal use of data. Step four regroups sale of data, transfer of data and grant of the right to use a database.

3.1 Introduction

This chapter aims to elucidate the anatomy of the 'data business' to set the scene for the tax law analysis that is conducted in the following chapters. To that end, first the different actors active in the data market, along with their functions within the market, are laid out. Second, the dynamics of the data market, arising from the interaction between the different actors are mentioned. Third, main issues identified in the functioning of the data market are explained. Finally, the taxable events related to the data business, which form the basis for the tax law analysis conducted in this book, are described.

X. Oberson, A. E. Yazıcıoğlu, *Taxation of Big Data*,
https://doi.org/10.1007/978-3-031-43372-6_3

The data business involves the use of systems of intelligence. Such systems combine "*software* (algorithms, business rules, machine-learning code, predictive analytics), *hardware* (servers, sensors, mobile devices, connectivity), *data* (contextualised and real-time), and *human input* (often judgment or questions)".[1] The current functioning of the data market is the result of the developments that occurred in software and hardware in the last decades. Prior to examine the anatomy of data business in detail, it is relevant to mention some general information relating to software and hardware and their substantial development.

In its narrow sense, software is computer code.[2] In its broadest sense, the term software encompasses computer code, data and knowledge represented in that data.[3] Once developed, software can be copied and shared without any significant costs. Software has the merit of being the first tool in the human history that can 'make itself', in more plain words that is 'able to learn'.[4] It learns how to recognise patterns and how to find hidden insights in the data.[5] Over the years, software has become increasingly tied to users, allowing a progressively high level of personalisation.[6] As a matter of fact, a switch has occurred from 'one-to-many' (where the users learned the system) to 'one-to-one' (where the system learns the user).[7] While early software required users to conform to it, 'new-machine-age' software conforms to users, like Alexa that self-configures to the user's patterns.[8] Software functions by providing a series of instructions directed at the processor.[9] It can implement inventions that are increasingly powerful in what they do, like finding the best driving routes.[10] Such smarter software becomes possible with the increasing power of computer systems.[11] Thus, development of smart software necessitates advances in hardware.

Hardware has gained substantial processing power since the 1960s. Golden Moore, one of the founders of Intel, stated in an article written in 1965 that the amount of integrated circuit computing power that could be bought for one dollar roughly doubled every year.[12] The time frame needed for the processing power to double was later amended to every 2 years. This observation is being referred to as 'Moore's Law'. Clearly, Moore's Law is not a 'law of physics' in the traditional sense, but rather a statement about the constant and successful efforts of computer

[1] Malcolm et al. (2017), p. 46.
[2] Meisel (2013), p. 3.
[3] Meisel (2013), p. 3.
[4] Malcolm et al. (2017), p. 46.
[5] Malcolm et al. (2017), p. 46.
[6] Meisel (2013), p. 15.
[7] Malcolm et al. (2017), p. 51.
[8] Malcolm et al. (2017), p. 61.
[9] Meisel (2013), p. 19.
[10] Meisel (2013), p. 13.
[11] Meisel (2013), p. 14.
[12] Brynjolfsson and McAfee (2014), pp. 40–41.

industry's engineers and scientists.[13] Such a sustained success, which is also due to the looser physical constraints in the digital world, can simply not be observed in other domains.[14] This success can be illustrated by means of comparing the progress of computer chips with cars. If a car from 1971 could have improved at the same rate as computer chips, the car models of 2015 would have had top speeds of about 420 million miles per hour.[15]

The Moore's Law effect on processors, memory, sensors and many other elements of computer hardware does more than just making computing devices faster, cheaper, smaller and lighter. It also allows these devices to effectuate things that previously seemed out of reach.[16] The fact that the chips got so small that they can be built into everything from kettles to cars is Moore's Law in motion. The cloud had the effect of 'turbocharging' Moore's Law, by enabling hyper-powerful computers to be tied together.[17] At some point Moore's Law must slow down, and that process may already have started.[18] Such a slowdown will, however, not necessarily translate into the collection of less data.

3.2 Different Actors Active in the 'Data Market'

The five main actors active in the data market are the following: (1) two or multi-sided platforms; (2) data brokers; (3) other data aggregators; (4) users; and (5) microworkers. This section briefly describes the most essential characteristics of each actor (or their business model) as well as the main role they play in the data market, without examining their impact on the overall market. The dynamics of the data market, as well as the main issues identified within that market, are analysed in the subsequent sections.

3.2.1 *Two or Multi-Sided Platforms*

Platforms operating in the data market are typically acting in a two or multi-sided manner. Two (or multi)-sided platforms are not specific to the data market. Banks, newspapers, television and shopping malls are the most commonly cited examples of such platforms.[19] A two (or multi)-sided market can best be described by contrast to

[13] Brynjolfsson and McAfee (2014), p. 41.
[14] Brynjolfsson and McAfee (2014), pp. 41–42.
[15] Cross (2017), p. 55.
[16] Brynjolfsson and McAfee (2014), p. 52.
[17] Malcolm et al. (2017), p. 47.
[18] See, for example, Brynjolfsson and McAfee (2014), p. 42; and Cross (2017), pp. 56–63.
[19] For more details on this point, see Collin and Colin (2013), p. 24.

a single-sided market. A single-sided market is characterised by the fact that sellers engage with only one specific set of customers, like a reader buying a book in a bookshop.[20] On the other hand, two (or multi)-sided platforms are characterised by the existence of two (or more) sets of agents interacting through an intermediary or a platform and by the fact that the decisions of each set of agents affect the outcomes for the other set(s) of agents, typically through an externality.[21] For example, newspaper and television bring together the audience and advertisers, shopping malls act as an intermediary between stores and consumers, banks take deposits from their retail banking customers and leverage those deposits to make loans or to invest in financial markets.

While entities active in the data market frequently follow a two or multi-sided business model, it is not possible to categorise an overall entity uniquely as a two (or multi)-sided platform. In fact, the same entity can manage a two or multi-sided platform in one market and act in a single-sided manner in another market. The most suitable example to illustrate this point is Amazon.com, which combines single-sided and two-sided selling strategies. For some products, such as books, Amazon.com buys the product at a wholesale price and sells it for a retail price: a typical single-sided model. For many other products, however, Amazon.com provides a web portal through which a producer commercialises its product by setting its retail price, and a consumer purchases that product directly by that producer.[22] Since an entity can simultaneously act in a single-sided, two-sided and multi-sided manner in different markets, the term 'two (or multi)-sided strategies' (or business models) seems more appropriate than the term 'two (or multi)-sided platforms'.[23] Nevertheless, this book refers to 'two (or multi)-sided platforms' for distinguishing such platforms from data brokers and other data aggregators and for the sake of simplicity. Hereinafter, the term 'multi-sided platforms' refers to entities following a two (or multi)-sided business model involving the collection of data.

The two characteristics of multi-sided platforms that are especially relevant for the examination of taxation of data are indirect network effects and non-neutral pricing strategies. *Indirect network effects* occur when an increase in end-users on one side of the platform increases the utility of end-users on another side of the platform.[24] For instance, within the framework of an online platform assisting individuals in renting accommodation by linking hosts and guests, both the hosts and the guests benefit if there are more end-users on the other side.[25]

---

[20] OECD (2018), § 48.
[21] Rysman (2009), p. 125.
[22] Rysman (2009), p. 130.
[23] Rysman (2009), p. 130.
[24] OECD (2018), § 50.
[25] OECD (2018), § 50.

In the context of multi-sided platforms, *non-neutral pricing strategies* are common, since the prices tend to reflect the effects of the externalities.[26] To the extent where one side produces a positive externality on another side (for example more clicks by users in the Google search engine on links sponsored by advertisers), the prices charged to the side benefitting from the positive externality can be increased (in our example, advertisers can be charged more heavily).[27] In a similar vein, prices below marginal cost[28] or even negative prices can easily arise within multi-sided platforms.[29] This would typically be the case when completing a transaction is beneficial to one side of the platform but costly to the other, so that the latter must be subsidised in order to transact voluntarily.[30] In more technical terms, prices below marginal cost or negative prices are due to 'usage externality', which is the benefit a user on one side generates for a user on the other by actually engaging in transactions on the platform.[31] To continue with the search engine example, users of the search engine may benefit from a zero or a negative price. Such subsidisation is intended to offset the negative externality produced by the other side, i.e., displays of intrusive advertisement, and to motivate more users to have recourse to the search engine.

3.2.2 Data Brokers

While multi-sided platforms seem to be perceived as the biggest data aggregators by the general public (probably due to social media and metaverse), the information that a data broker has on consumers can be as substantial as (and in some cases more substantial than) the information retained by a multi-sided platform. For example, Acxiom Corporation is reported to have more than 23,000 computer servers that are collecting, collating and analysing consumer data. The servers of the company are known to process more than 50 trillion data transactions a year. Company executives have indicated that its database contains information about 500 million active consumers worldwide with about 1500 data points per person.[32]

"Data brokers acquire a vast array of detailed and specific information about consumers; analyse it to make inferences about consumers, some of which may be considered sensitive; and share the information with clients in a range of industries.

---

[26] Externality can be described as the positive or negative effect produced by someone's decision on someone else, despite the lack of any specific contractual arrangements between the two parties concerned. [See, for example, Caldari and Masini (2011), p. 716].

[27] On this issue, see, for example, Collin and Colin (2013), p. 23.

[28] Marginal cost can be defined as the increase in total cost that arises from an extra unit of production. [See, for example, Mankiw (2017), p. 256].

[29] Rysman (2009), p. 130.

[30] Hermalin and Katz (2018), p. 119.

[31] Hermalin and Katz (2018), p. 113.

[32] Singer (2012).

All of this activity takes place behind the scenes, *without consumers' knowledge*".[33] Data broker industry constitutes a lucrative industry, generating approximately USD 200 billion on an annual basis.[34] For an overview of the business model traditionally employed by data brokers, see Fig. 3.1.

Data brokers may gather their data from commercial, governmental and other publicly available sources.[35] Data brokers frequently obtain data from each other as well, rather than directly from an original resource.[36] Data collected by data brokers are not only limited to online data, but also include offline data.[37] Collection of offline data by data brokers is typically achieved via placement of cookies on consumers' browsers. Data brokers may also purchase offline data relating to consumers from stores and companies that the consumers interact with in the course of their daily activities. Data collected by a broker may include, for example, purchase history, warranty information, credit card information, driver's license information, age, gender, data about real estate owned, income, education and occupation.

While technically data brokers may obtain their data directly from consumers with their consent, this is usually not the case. Consumers are largely unaware of the fact that data brokers are collecting and using their information.[38] As a matter of fact, data brokers are predominantly proceeding to secondary or tertiary uses of information gathered in the ordinary conduct of business and governmental activities. Secondary (or tertiary—depending on the number of necessary transactions to obtain the data) use of information can be described as any use of information beyond the purpose for which the data were originally gathered.[39] The information gathering effectuated by stores, companies or governmental agencies in the course of their direct interactions with consumers corresponds to the primary use of information.[40] Subsequent transfer or sale of that information by the store, the company or the governmental agency concerned to the data broker forms a secondary use of information. It is extremely cumbersome, if not impossible, for a consumer to be able to pinpoint the secondary and tertiary uses of their information by a data broker. Even when consumers may identify such uses, it would be impossible for them to determine how the data broker obtained their data. Consumers may have to retrace the path of data through a series of data brokers to obtain that information.[41]

To develop their products, data brokers use the raw data they collected along with the inferred data they derive. For instance, a data broker may infer that an individual

[33] The US Federal Trade Commission (2014), p. vii. Emphasis added by the authors.

[34] Wlosik (2022).

[35] The US Federal Trade Commission (2014), p. iv.

[36] The US Federal Trade Commission (2014), p. iv.

[37] The US Federal Trade Commission (2014), p. v.

[38] The US Federal Trade Commission (2014), p. iv.

[39] Laudon (1996), p. 99.

[40] Laudon (1996), p. 99.

[41] The US Federal Trade Commission (2014), p. iv.

Fig. 3.1 Business model of data brokers

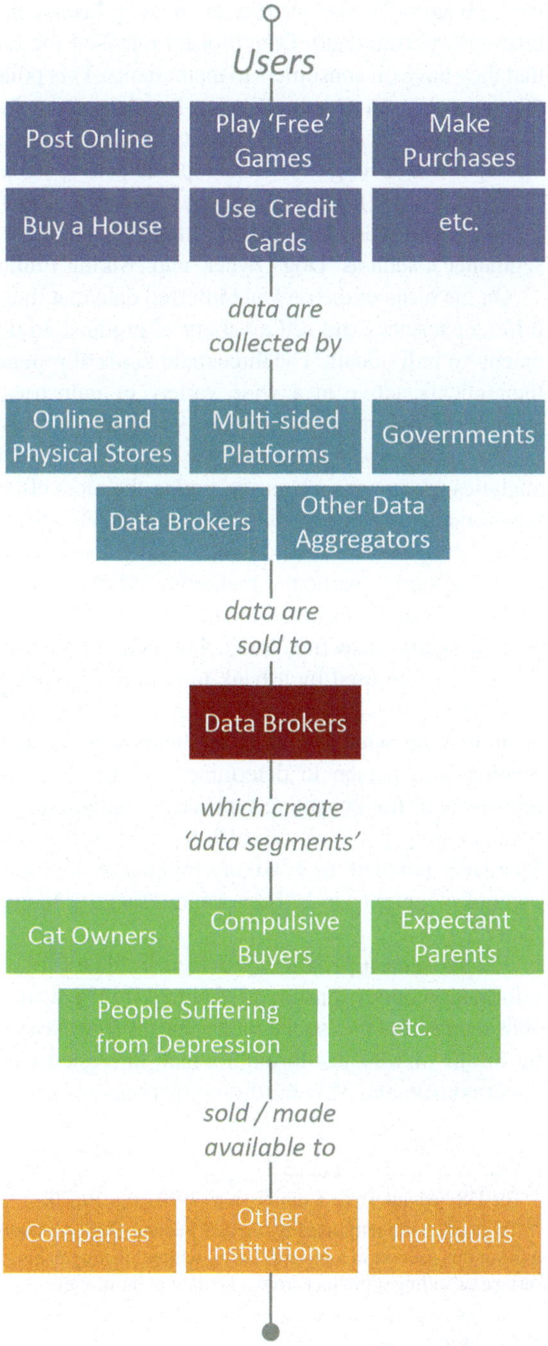

with a boating license (having the boating license is the raw data) has an interest in boating (inferred data). Data brokers may sell the raw data and/or the inferred data that they have on consumers to their clients. This point can be illustrated by means of the example of a data broker that has data about Jane Doe living in London. The broker may sell raw data elements about Jane, including her name, age, marital status, interests (e.g., dogs and participation to bike&hike events) and address (123 Main Street). The broker may also use this raw data, produce inferred data on this basis and place Jane Doe in different categories (frequently referred to as 'data segments'), such as 'Dog Owner' and 'Biking Enthusiast'.[42]

On the basis of the raw and inferred data that they have, data brokers may render different services and sell a variety of products to other companies, or -to a limited extent- to individuals. The three main kinds of products prepared by data brokers for their clients active in a wide variety of industries are the following: marketing products,[43] risk mitigation products and people search products.[44]

Marketing products include sale of raw data (like e-mail addresses of consumers), analytics products (such as suggesting the most efficient media channel to advertise the product, establishing data segments and effectuating targeting) and establishment of marketing scores (for example ranking consumers on the basis of how likely they are to respond to particular marketing efforts or to make a purchase).[45]

Risk mitigation products are mainly used for verifying the identities of the clients' customers and detecting fraud. For instance, a data broker's identity verification product can be used by a bank to confirm that Jane Doe at 123 Main Street, who wishes to open a bank account, is in fact Jane Doe.[46] Calculating the credit score of an individual who makes an application for a loan or verifying the information relating to a person to determine whether they can receive lower life-insurance premiums or not constitute typical risk mitigation services.[47] Segments established for marketing purposes can also be used in risk mitigation services. For example, Jane Doe 'labelled' as a 'Biking Enthusiast' can suffer from this qualification in an insurance context, since 'biking enthusiasm' is likely to be interpreted as a high-risk factor.

People search products provided by data brokers generally consist of offering information about consumers obtained from governmental and other publicly available sources, such as social media sites. People search services are typically rendered by means of websites through which interested parties can search for information about consumers. This third type of product offered by data brokers diverges to a

---

[42] The US Federal Trade Commission (2014), pp. 19–20.

[43] 'Product' is the term used by the US Federal Trade Commission in its report. The term is being used in this section in a general sense, without distinguishing whether a data broker is rendering a service or selling a product from a tax law point of view.

[44] The US Federal Trade Commission (2014), p. ii.

[45] The US Federal Trade Commission (2014), pp. 23–31.

[46] The US Federal Trade Commission (2014), pp. 32–34.

[47] Wlosik (2022).

certain extent from the first two, in that it is mainly intended to be employed by individuals, and not by other companies.[48]

3.2.3 *Other Data Aggregators*

Aside from multi-sided platforms and data brokers, there are some other data aggregators that collect and use consumer data with the main aim of better marketing their own products.[49] For example, a store might collect data simply to know its customers better and gain insight into their preferences, like whether and to what extent they are likely to respond to sales. As already mentioned (in Sect. 3.2.2), such aggregators may sell the data they collect to data brokers as well. Governments and public entities also collect data and constitute 'data aggregators' in the broad sense of the concept. For the purposes of this book, the term 'data aggregators' refers to entities commercialising data. The term is, thus, intended to primarily cover private entities.

3.2.4 *Users*

In the context of a transaction, the 'user' is typically identified as the buyer or consumer of a good or service, regardless of whether the transaction takes place physically (in person) or virtually (online).[50] For the purposes of multi-sided platforms, two parties utilising the platform, such as a buyer and a seller, are both considered as users of the services provided by the platform.[51] Throughout this book, the term 'user' is employed in its most traditional sense, i.e., buyer or consumer of a good or service.

Users form the *sine qua non* component of the data market. The significance of the role of users, which cannot be overstated, has been emphasised on a number of occasions by different politicians and academics. For instance, in March 2009, Meglena Kuneva, European Commissioner for Consumer Protection at date, stated that "personal data is the new oil of the Internet and the new currency of the digital world".[52] Posner and Weyl describe users as 'a vital cog in the information economy'.[53] In the same vein, Thimmesch indicates that consumer information is the

---

[48] The US Federal Trade Commission (2014), pp. 23–35.

[49] Thimmesch (2016), p. 5.

[50] Aslam and Shah (2020), p. 8.

[51] Aslam and Shah (2020), p. 8.

[52] World Economic Forum (2011), p. 5.

[53] Posner and Weyl (2019), p. 71.

'fuel of the big data era'.[54] In a much direct manner, Purtova asserts that 'the most vital resource for the Information Industry is the user'.[55]

Prior to explaining the significance of users for the data market from a scientific point of view, it is important to underline the reason for which users are considered to be absolutely essential, to the point to be qualified as 'the vital cog', for the information economy. The reason is extremely simple and straightforward: *the data market cannot exist without users*. Multi-sided platforms, data brokers, other data aggregators as well as microworkers all depend on, and exist because of, users. If users are taken out of the equation, there is no data market. This fact seems simple and obvious, but it is not emphasised enough and seems to be frequently underestimated. This very fact forms the backbone of the analysis conducted in this book. No matter the main subjects of the current discussions in different fields, the data market is not predominantly about data aggregators, or governments, it is about users. Any viable policy suggestion should take this fact (and it is a fact, not an opinion) into consideration.

Without a certain mass of users, multi-sided platforms can simply not exist. The rationale behind this fact begins with *direct network effects*. As explained by the OECD, "in digital markets, utility from the consumption of a specific good or service is often dependent on the number of other end-users consuming the same good or service."[56] Social media and online messaging services constitute typical examples.[57] A user wishing to employ a social media platform or an online messaging service will generally opt for the platform or the service that has a large number of users. Will also reveal to be important, the choice of platform or service that was already made by the user's friends, family and colleagues in a predominant manner. Direct network effects thus generate a positive feedback loop: the more users adopt the service, the more valuable it becomes to potential adopters.[58] This positive feedback loop also works in reverse: should the adoption fail to reach a critical mass of users, the good or service may fall into a 'death spiral' and ultimately disappear.[59] Undisputedly, a multi-sided platform that cannot attract a critical mass of users on one side, can simply not provide services to customers positioned on other sides. By means of example, a social media platform that does not have enough users will hardly be an attractive business partner for advertisement purposes.

Without the data of users, data brokers cannot exist. Data brokers are in the business of collecting, aggregating and analysing data of users and selling them to interested parties. Should users' data become unavailable for one reason or another, the whole data brokerage industry would cease to exist. Other data aggregators'

[54] Thimmesch (2016), p. 2.

[55] Purtova (2015), p. 104.

[56] OECD (2018), § 42.

[57] OECD (2018), § 42.

[58] Varian (2018), p. 227.

[59] Varian (2018), p. 227.

primary activities are conducted in markets other than the data market, and their business models are not mainly built on users' data. They would, thus, continue to sell their products and commercialise their services even in the absence of user data, albeit suffering from the negative consequences of not being able to employ the data concerned.

For the data market to be able to function, user data are needed constantly and in consequential amounts. In fact, the data collected are not utilised only to sell advertisement spots to a variety of companies or to provide risk mitigation services to insurance companies, banks and similar institutions. Data are also necessary to train Artificial Intelligence (AI) and enable machine learning (ML). Moreover, they empower the companies to 'learn by doing', which allows to lower the per unit cost or improve the quality of a good or service over time.[60]

3.2.5 Microworkers

> If customer data is the new oil, the people doing ghost work operate the rigs.[61]

Microwork, which is an alternative term for 'crowdwork', refers to the carrying out of tasks that necessitate human computation and, for that very reason, that cannot be conducted by the AI. In fact, data alone are insufficient to train a neural net. The process of actually training the net requires huge numbers of computations.[62] This requirement gives rise to the necessity of 'microwork'. For example, there is currently a mismatch between Facebook's needs and the reasons for which users post photographs. Since they expect that their friends would understand the context of their photographs, users tend to provide little information accompanying the images they post. This tendency causes the data to be of 'low-quality' for Facebook, which tries to nudge users to provide useful labels by inducing them to write comments explaining photographs or to associate emotions with them. What Facebook really needs is the capacity to ask users simple questions about the photographs and receive answers from them.[63] Microworkers can typically intervene at this stage: they can 'tag' the photographs and/or associate images with emotions.

Microworkers can work alongside software to deliver seemingly automated services to consumers.[64] An example given by Gray and Suri illustrates the phenomenon in a clear manner. Emily, who lives in Chicago, requests a ride through the Uber app on her smartphone and an Uber driver, Sam, responds. Neither Emily nor Sam is aware of the fact that Uber's software just flagged Sam's account, because Uber's Real-Time ID Check could not match the selfie that he took this morning

60 Varian (2018), p. 229.
61 Gray and Suri (2019), p. 41.
62 Posner and Weyl (2019), p. 74.
63 Posner and Weyl (2019), pp. 74–75.
64 Gray and Suri (2019), p. 7.

with his photograph on record. Ayesha, who lives in India, is a microworker accepting tasks from an Indian platform taking requests from Uber. She accepts a job routed from Uber to the Indian platform's software, accesses Sam's two photographs, compares them and determines, in a limited time frame allocated to her by the Indian platform, whether the two photographs belong to the same person. As Ayesha clicks 'yes', Sam's account is authorised to pick Emily up.[65] While Emily and Sam continue to assume that they are using a completely automated platform, Ayesha moves on to her next task.

While almost all microwork tasks are time-sensitive, microworkers do not intervene exclusively in the context of real-time user services. In fact, a non-negligible part of the tasks completed by microworkers is destined to train the AI. Amending translations, tagging photographs (with emotions or with the names of the items that are on the photographs, like a table) as well as reading texts in distorted scans and typing them in a box form typical examples of microworking tasks destined to train the AI.

Microwork is conducted via microtask platforms, such as Amazon.com's Mechanical Turk (MTurk) and Microsoft's internal Universal Human Relevance System (UHRS). Access to tasks is usually granted once a microworker registers and their account is approved. Microwork is conducted on a task-by-task basis, in accordance with detailed instructions prepared by requesters. Generally, microworkers and requesters do not interact in a direct manner. The identity of the microworker is normally not even disclosed to the requester, and vice versa. Platforms tend to provide microworkers with a 'worker ID', which is a series of random letters and numbers unique to the microworker concerned. In the same vein, requesters using such platforms designate a nickname.

Microworkers seem to be the true invisible actors of the data market.[66] Requesters using microtask platforms neither have any information relating to the identity of microworkers, nor do they establish any communication with them. They post a task, an 'ID number' accepts it and the work is submitted within the designated time frame. Users, who think they are using automated services or products, generally ignore the very existence of microworkers. Gray and Suri argue that the human labour powering many mobile apps, websites and AI systems is often intentionally hidden from users.[67] They coined the term 'ghost work' to refer to this 'opaque world of employment'.[68]

---

[65] Gray and Suri (2019), pp. 12–13.

[66] While microworkers are not exclusively relevant for the purposes of the data market, the scope of this book is limited to the market concerned.

[67] Gray and Suri (2019), p. 7.

[68] Gray and Suri (2019), p. 7.

3.3 Dynamics of the 'Data Market'

Identification of the main actors of the data market is the first step towards understanding the functioning of the market concerned. The second step consists of determining the main dynamics. Clearly, examining all the major dynamics of the data market does not fall within the scope of this book. Are only examined the current features of the market that have a significant effect on the tax law analysis that is conducted in this contribution.

The term 'dynamics' is used, instead of 'characteristics' or 'features', given that the aspects examined in this section are not intrinsic characteristics of the data market. They constitute the current features that may be subject to change in the future. The primary purpose of examining the chosen dynamics of the data market is to set the scene for the description of the fundamental issues identified by the authors in the functioning of the data market.[69]

The main dynamics of the data market are the following: (1) users expect the services delivered over the Internet to be free; (2) users are not always obtaining free services in exchange for their data; (3) the data market is dominated by a small number of companies; and (4) working conditions of microworkers need to be ameliorated.

3.3.1 Expectation of Free Digital Services

The general public seems to have developed a strong attachment to the idea that services, information and software provided through the Internet should be free of charge. This attachment has been formed over time, due to several reasons.

First should be mentioned, the different options entertained during the designing process of the Internet. Many early designs of the Internet supported technologies that would have allowed receivers of information to automatically pay the providers.[70] The designs concerned made use of two-way links, where every piece of information would effectively carry its full provenance with it.[71] For example, the pre-Internet Minitel system of France incorporated a structure enabling micropayments.[72] However, what eventually became the mainstream Internet was a collaborative platform within government, military and academic circles where participants were assumed to be interested in collaboration for reasons external to

[69] See Sect. 3.4.

[70] Posner and Weyl (2019), p. 71.

[71] Posner and Weyl (2019), p. 71.

[72] Posner and Weyl (2019), p. 71.

commercial and economic motivations.[73] Emphasis was placed on lowering barriers to participation.[74]

During the 1990s, Internet companies relentlessly pursued users under the banner 'usage, revenues later'.[75] The network effects created by this strategy encouraged many venture capitalists to adopt it and enlarge their user base, albeit having an unclear business model.[76] During this extended period of free access to services, the general public developed a strong attachment to the idea of completely free services.[77] As a matter of fact, individuals perceive the services delivered over the Internet, like Google's search engine, as a natural part of the Web.[78]

Some companies did contemplate charging their users for the services they provide. For instance, Google initially considered implementing user fees and paid subscriptions.[79] This strategy, which would have gone against the general public tendency, could not be adopted for two other reasons. First, many online services provided at that time were occasional and small, with the result that investment in the development of infrastructure that would have been needed to keep track of payments was not cost-justified.[80] Second, the presence of young hackers who were determined to preserve the 'freedom' of the Internet made charging for anything, including established forms of intellectual property such as music, challenging.[81]

These different forces in the industry established an environment where users expected digital services to be free of charge, and therefore were reluctant to pay for anything. Quite understandably, companies active in the sector searched for a way to monetise their activities. First Google, then Facebook turned to targeted advertisement by means of utilising user data.[82] Hence began the era, during which users, who may not even be aware of the productive value of their data, became accustomed to surrendering their data in exchange for 'free' services.[83]

---

[73] Posner and Weyl (2019), p. 71.

[74] Posner and Weyl (2019), p. 71.

[75] Posner and Weyl (2019), p. 71.

[76] Posner and Weyl (2019), p. 71.

[77] Posner and Weyl (2019), p. 71.

[78] Meisel (2013), p. 6.

[79] Posner and Weyl (2019), p. 71.

[80] Posner and Weyl (2019), p. 71.

[81] Posner and Weyl (2019), p. 72.

[82] Arrieta-Ibarra et al. (2018), p. 41.

[83] Arrieta-Ibarra et al. (2018), p. 41.

3.3.2 *Users Do Not Always 'Get to' Exchange Their Data with Free Services*

Proponents of the current dynamics of the data market systematically underline the fact that users are exchanging their data with free digital services, which have a substantial value. The existence of a consumer surplus[84] produced by the services concerned is beyond discussion. In fact, according to a study conducted by Brynjolfsson and Oh, covering the years between 2007 and 2011, the consumer surplus deriving from the free digital services amounted to USD 106 billion in the US alone.[85] That being said, in the current functioning of the data market, users are not always getting free services in exchange for their data.

First, users are generally not even aware of the fact that their data are being collected and commercialised by data brokers. On this very basis, it is clear that no free services are being provided to users by such brokers. While the products and services commercialised by the latter may produce some indirect benefits for the former, such products and services are simultaneously provoking significant risks. For instance, risk mitigation services and products provided by data brokers may help prevent fraudsters from impersonating unsuspecting users and thereby benefit users.[86] In a similar vein, online advertising is exceedingly perceived as being a consumer service, since it is thought to provide the right information to the right person at the right time.[87] On the other hand, different applications (like a loan application) made by consumers may be denied due to a piece of information contained in a risk mitigation product commercialised by a data broker.[88] Consumers have limited means to correct such information, which may affect their daily lives in a negative manner. Furthermore, some data segments created by data brokers can contain sensitive information about users, such as 'Diabetes Interest', 'Depressed Young Adult' and 'African-American Professional'.[89] Finally, the fact that data brokers tend to keep the data they have for a long time, may create security risks for consumers whose data are being stored.[90]

Second, users may not realise the amount of data that are being collected by different equipment and devices they are using on a daily basis. While a smartphone or a smartwatch is expected to collect a certain amount of data; a refrigerator, a coffee machine, a car or an airplane may not necessarily be perceived as data collecting devices by users. The data collected by these everyday items may produce

---

[84] Consumer surplus can be defined as the maximum amount that a buyer is willing to pay for a good or service minus the amount that the buyer actually pays for it [See, for example, Mankiw (2017), pp. 134–138].

[85] Brynjolfsson and Oh (2012), p. 3.

[86] The US Federal Trade Commission (2014), p. v.

[87] Collin and Colin (2013), p. 15.

[88] The US Federal Trade Commission (2014), p. v.

[89] The US Federal Trade Commission (2014), pp. vi and 21.

[90] The US Federal Trade Commission (2014), p. vi.

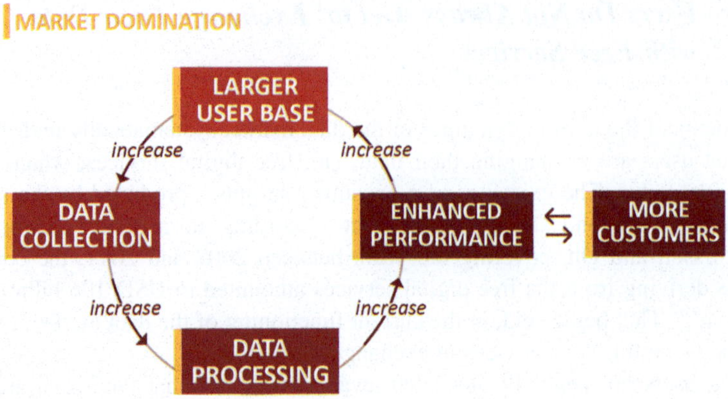

Fig. 3.2 Market domination

a positive impact on users, especially through the amelioration of the future models of such devices, but this impact is too indirect to be considered as a 'free service in exchange of data'.

Finally, some subscription-based or otherwise paid services may still (and generally do) collect user data. For instance, the Netflix data warehouse stores about 10 petabytes of information. The company does not only collect the data about the most popular contents, the ratings, the most common searches and other similar information. All the event data flowing from devices and applications, websites and anything that consumers use to interact with the Netflix platform are collected.[91] In the same vein, during a typical Uber ride, the following data are captured: data on request, location, time, route taken, device used, payment and tip, driver, passenger, driver rating and passenger rating.[92]

3.3.3 The Data Market Is Dominated by a Small Number of Companies

The fact that the data market is dominated by a small number of multi-sided platforms as well as a small number of microtask platforms gives rise to a number of issues that are pertinent for the purposes of the tax law analysis conducted in this book. The market dominance of certain data brokers (to the extent where such a dominance exists) is not relevant in the context of this contribution since data brokers do not directly interact with users or microworkers.

[91] According to Daniel C. Weeks, see Weeks (2015). See also, Malcolm et al. (2017), pp. 58–59; and Vanderbilt (2013).

[92] Malcolm et al. (2017), p. 47.

As illustrated in Fig. 3.2, multi-sided platforms' market domination is based on four key factors, which produce an impact on each other. The term that best describes such dominance would be 'vicious cycle' since that dominance creates negative effects on the overall market. The 'cycle' works in the following manner: (1) the market-dominating multi-sided platforms manage to gather the majority of users due to network effects; (2) the substantial number of users they have enables them to collect enormous amounts of data (more data than that can be collected by other platforms active in the same sector—to the extent where such platforms can 'survive'—); (3) the massive amounts of data collected allow them to process such data and derive important benefits (including training their AI); (4) the benefits derived from the processed data enable the platforms to produce better performing products and/or services for their several 'sides' (e.g., provision of better online targeted advertisement and better user experience); (5) which, in turn, results in more users to opt for the platforms concerned.

3.3.3.1 Larger User Base Leading to More Data Collection and More Data Processing

Users may opt for utilising the services of a multi-sided platform due to *network effects*. For example, a user who wishes to use an online messaging service would typically opt for the service employed by the majority of their family, friends and colleagues (direct network effects). Some services offered by multi-sided platforms may not present traditional network effects. Search engines form the most typical example of such services. What predominantly matters from a user perspective is the quality of the search engine's performance, and not the number of people employing it.[93] That being said, the number of people using the engine may, nonetheless, form a strong indicator of better performance in users' minds. In the context of search engines, supply side effect seems to be in play.[94] More users lead to more data, which lead to more product improvements, which, in turn, lead to more users.

In any case, one or two multi-sided platforms manage(s) to attract *the majority of the users* and obtain, thereby, a certain market dominance. While the terms 'monopoly' and 'oligopoly' are being used to describe this situation, the authors opt for the term 'market dominance', due to the fact that the exact meaning of 'monopoly', and respectively of 'oligopoly', may not actually match the phenomenon, depending on the circumstances. Bearing that in mind, it is possible to state, in general terms, that multi-sided platforms benefit from an 'oligopolistic' effect on the user base.

A large user base inevitably results in the *collection of more substantial user data*, which in turn leads to *the ability to process more data*. Clearly, being able to collect more data does not automatically translate into being able to process more data. Multi-sided platforms should also have the relevant tools to proceed with data

[93] Varian (2018), p. 230.

[94] Varian (2018), p. 232.

processing, which requires a substantial investment and serious know-how. In other words, to be able to benefit from their user base, the platforms should invest in data analysis. Nevertheless, a platform with a large user base that was able to develop the relevant AI and processing skills, benefits from a dominant position in the segments of data collection and data processing.

When considered *from a labour perspective*, it is commonly stated that the dominant position in data generation/collection and processing stages give rise to 'monopsony' or 'oligopsony'. Again, using the term 'market dominance' seems to be more appropriate in this context. That being said, between 'monopsony' and 'oligopsony', it would be more suitable to opt for 'oligopsony' both in data generation/collection and processing stages. The two stages need to be distinguished from each other while explaining this oligopsonistic effect.

Some contributors argue that data should not be treated as capital (as they currently are) but as labour.[95] Should data generation be perceived as labour, it would be recognised that users have the possession of their data and the pay-offs of that data would be channelled to individual users, instead of companies.[96] Data generation would thus be considered as legitimate work, rather than as a legitimate side-effect of consumption. Should such a view effectively be followed, the fact that the data are predominantly collected by a few companies is likely to lead to oligopsony in the 'data generation market'. User-labourers will only have an extremely limited number of companies to work for, which will be able to determine to a great extent the conditions under which the user-labourers will provide their work (unless specific legislation is enacted). Currently, this oligopsonistic effect translates in the maintenance of 'data as capital' equilibrium, where users are not even aware of the value of the data they are creating on a daily basis for the market dominant multi-sided platforms, which can continue to benefit from the free or extremely cheap availability of data.[97]

In the data processing stage, the oligopsonistic power is especially relevant from a microworker perspective. For these workers who intervene through the completion of different tasks during the data processing stage, such oligopsonistic effect implies that their working conditions can be determined to a great extent by a limited number of companies that are processing the majority of the data. The market-dominant platforms can set the norms for the price to be paid upon the completion of a task, the time frame during which the tasks need to be completed, the details provided in the instructions etc. In the current standing of the microtasking industry, none of these details seem to be determined with the welfare of the microworkers in mind. This oligopsonistic effect is also strengthened by the existence of a limited number of microtask platforms, which do not seem to intervene in view of creating a balanced microtask market. The working conditions of microworkers are examined in the following section, 3.3.4.

[95] See notably Arrieta-Ibarra et al. (2018) on this point.

[96] Arrieta-Ibarra et al. (2018), pp. 39–40.

[97] Arrieta-Ibarra et al. (2018), p. 41.

3.3.3.2 More Data Collection and Processing Leading to Enchanced Performance, More Customers and an Even-More-Enlarged User Base

The collection and processing of more data enable the companies to render *better services* or to produce *better products* not only for the user-side of the platform, but also for the other side(s) of the platform. The vast amounts of data that are gathered and turned into information, knowledge and understanding empower the platform to 'learn by doing'.[98] Learning by doing can be described as the experience gained by a company over time, which results in lower costs and/or higher quality per unit,[99] both of which benefit to the company as well as to its users and clients. For instance, from a user perspective, learning by doing may allow to ameliorate user interface design and to render better-performing social media services; and from a customer perspective, learning by doing may allow the platform to render extremely accurate online targeted advertisement services.

The effect produced by learning by doing inevitably results in *attracting more users*, along with *more customers* on the other sides of the platform. Users are attracted to the platform, since they benefit from an easier-to-use and more efficient service and customers prefer the market-dominant platform due to the quality of its services and/or products, as well as its large user base (indirect network effects). For instance, users certainly opt for Facebook due to direct network effects as well as the quality of the service provided by the platform. The more Facebook becomes dominant in the market, the more it solidifies its user base and the more data it collects and processes to render its services to the other sides of the platform. It is, therefore, not surprising that Facebook can provide extremely efficient online targeted advertisement services and controls a significant part of the digital advertisement sector, alongside Google.[100]

3.3.3.3 Perpetuating the Vicious Cycle

With the fourth and last step of reaching an even-more-enlarged user base, the cycle is completed. An even-larger user base, partially due to better-performing services, continues to set in motion the dynamics that 'perpetuate the vicious cycle'. The oligopolistic and the oligopsonistic effects giving rise to the vicious cycle, which are referred to as market dominance throughout this book, produce a significant impact over the tax law analysis.

As a result of the dynamics of the 'vicious cycle', an extremely limited number of multi-sided platforms dominate a number of markets. One key example of such markets is online targeted advertisement. Galloway summarises this phenomenon as

---

[98] Varian (2018), p. 230.

[99] Varian (2018), p. 229.

[100] Galloway (2017), pp. 107–109.

follows: "Advertisers have just two platforms to market their products online. Advertisers need Facebook much more than Facebook needs any one, or thousand, advertisers. If Ford Motor announced they were [. . .] no longer going to advertise on Facebook, the Detroit firm would be commended for their actions in the media, and the market would promptly cut 10 percent off their share price".[101]

The vicious cycle has been explained by Lanier in a unique way. The contributor coined the term 'siren server' to refer to market-dominant multi-sided platforms. He describes a 'siren server' as an elite computer, or coordinated collection of computers, on a network. "Siren Servers gather data from the network, often without having to pay for it. The data is analysed using the most powerful available computers, run by the very best available technical people. The results of the analysis are kept secret, but are used to manipulate the rest of the world to advantage. That plan will always eventually backfire, because the rest of the world cannot indefinitely absorb the increased risk, cost and waste dispersed by a Siren Server".[102]

3.3.4 Working Conditions of Microworkers Need To Be Ameliorated

As already mentioned, microwork is conducted via microtask platforms, such as Amazon.com's MTurk and Microsoft's internal UHRS. MTurk seems to be the most popular online microtask platform, which functions by allowing requesters to post 'tasks' that microworkers can complete for pay.[103]

Some of the main issues of the current microwork market that were identified by Gray and Suri in their book, *Ghost Work*, can be summarised as follows:[104]

- Microwork can be challenging to manage for requesters. Since microworkers remain anonymous in the system and tasks are granted on a first-come-first-served basis, a requester does generally not have the opportunity to work with the same microworker on several tasks. This results in a lack of understanding of the company's culture on the side of microworkers (who complete a variety of tasks for a variety of companies) and a need for requesters to constantly micromanage the workers.
- Microworkers need to work without breaks and be constantly online to be able to earn any decent amount of money in exchange for their work. This lack of flexibility becomes even more pronounced as a result of 'artificial time squeezes' imposed by requesters.

[101] Galloway (2017), p. 108.

[102] Lanier (2013), pp. 65–66.

[103] Dube et al. (2018), p. 2.

[104] For more details on this point, see, Gray and Suri (2019), Part II (Demanding Work), pp. 84–130.

- Microworkers do not receive any clear guidance or satisfying feedback. They do not have the possibility to contact someone on the requester side, who can answer to their questions relating to the tasks. Requesters are even allowed to modify the descriptions of tasks already accepted by microworkers. Should a requester deem that the work conducted by a microworker is of 'low quality', not only the microworker is not paid for the task concerned but also their score, which allows them to accept higher paying tasks, plummets, which condemns them to take on low-paying tasks for a certain period of time.
- To be able to take on tasks, microworkers must have their own appropriate software and devices. Requesters are not providing microworkers with any equipment necessary for the fulfilment of the tasks.
- Microworkers do not have the legal status of 'employee'. They are neither employed by the microtask platforms nor by the requesters. Quite the contrary, microtask platforms deem microworkers to be 'customers'. This category of workers can, thus, not benefit from any legal protection and remedies that are normally available to the majority of other workers.
- Microworkers frequently suffer from technical failures and glitches in the system. Their account can be temporarily (or permanently) suspended and their payments may not be effectuated in time (or ever).

On top of all these different issues, microworkers do not earn substantial amounts in return for their hard work. The top earners in the MTurk platform earn a little more than USD 7.25 per hour.[105] It must be admitted that, if the authors of this book were working on microtasks, they would probably not have earned USD 7.25 per hour, at least initially. One of the authors (Yazıcıoğlu) attempted to complete one task on MTurk. The remuneration for the task was USD 0.1. The author read the guidelines for 10 min and found them quite difficult to understand. In the meanwhile, the task was already taken by another microworker.

As the detailed research conducted by Gray and Suri clearly demonstrates, while the general information given about microwork may create the impression that the tasks are destined to be fulfilled by 'low-skilled' individuals, it is not the case. It is not surprising that approximately 70% of microworkers, active on different platforms, have university degrees.[106] When described (tagging photos, identifying cats and dogs, etc.), the tasks may seem overly simple, but in reality, an individual needs a decent level of English and good computer skills to be able to complete such tasks. Yet, the working conditions as well as the remuneration offered are even more precarious than those of low-skilled jobs.

The existence of such conditions may be explained by the significant market dominance of requesters. For example, according to a study conducted by Dube

[105] Gray and Suri (2019), p. 38.

[106] See Gray and Suri (2019).

et al., much of the surplus created by the microtask platform MTurk is captured by requesters, while workers are paid less than 20% of their productivity.[107]

3.3.5 Intermediary Conclusion: Tax Law May Be an Efficient Tool, but Is Not the Panacea

The current functioning of the data market is not ideal. A number of legislative solutions, especially from the perspectives of data protection law, competition law and labour law, would be welcome. With that in mind, the tax law suggestions made by the authors are not intended to fundamentally change the current market dynamics. Quite the opposite, the suggestions are designed to fit well within the current scheme of things, to be easy-to-implement and to be put in motion in a rather swift manner.

Many reasons justify such an approach. First, tax law, which becomes applicable only at the end of the chain and when the main problem has already occurred, does not constitute the best tool to solve most (if not any) of the issues mentioned in this section. Second, as examined below in Chap. 4, there seems to exist some technological limits to a number of the remedies that could potentially be implemented. Third, the fundamental change of the current system, which may be desirable in the long term, necessitates a robust international collaboration and a significant reform, which will inevitably take a considerable time to implement. The time needed to reform the entire data market should not preclude the competent authorities from making minor reforms in the tax system to allow the creation of a more sustainable data market, especially from a user and a microworker perspective, in the short and medium terms. Last but not least, even solely reforming the existing tax system in a fundamental manner can reveal to be highly challenging, as demonstrated by the developments that occurred in the last decade regarding international taxation of the digital economy.[108] It seems, therefore, more reasonable to concentrate on 'minor touches' rather than aiming to implement ambitious reforms. Such an approach is also justified, given the fact that very efficient 'minor touches' do exist and can solve the tax law problems to a great extent.

The suggestions made in this book are destined to respond to the immediate problems while taking the bigger picture into consideration, within the reasonable limits and in an efficient manner. While suggesting a panacea would certainly be

[107] Dube et al. (2018), p. 15.

[108] From a terminological perspective, the 'digitalisation of the economy' is more appropriate than the 'digital economy'. As underlined by many commentators, as well as the OECD, "[a]ll traditional businesses are digital, at least to some extent, and all digital businesses are traditional, at least to some extent" (Morton 2014). The use of the term 'digital economy' was adopted in this book not due to a terminological preference but for ensuring the ease of reading.

more desirable, it encounters one significant obstacle in the particular context of taxation of data: such a solution does not exist, at least not at this point in time.

3.4 Main Issues Identified in the Functioning of the 'Data Market'

In the light of the analysis conducted above and in the authors' opinion, the four following issues existent in the data market are especially relevant from a tax law perspective: (1) it must be acknowledged that raw data are not valueless; (2) the current functioning of the data market causes privacy concerns; (3) a non-negligible number of jobs are disappearing without any viable replacement; and (4) operational local scale can be reached without local mass.

3.4.1 *From 'It's Free and Always Will Be' to 'It's Quick and Easy':[109] Raw Data Are Not Valueless*

Any person who decides to effectuate some research about the data market will inevitably come across one of the two following 'observations' within the first 5 min: 'raw data are valueless' and 'data do not have any value until they have been processed'. This common, and yet clearly erroneous, 'belief' seems to constitute one of the biggest problems of the data market.

Data are the primary raw material of the fourth industrial revolution, just as coal was the primary raw material of the first industrial revolution and oil, steel and electricity were the 'new' raw materials of the second industrial revolution.[110] This 'new' primary raw material is processed by 'the new machines', which consist of systems of intelligence combining hardware, AI, software, data and human input to create value aligned to a specific business process or customer experience. The end result of that process is being used in the typical 'new' business models, which are commercial models that monetise services and solutions based on systems of intelligence.[111]

Raw data are, in fact, not very different from other 'raw materials', which become more valuable when they are processed. It does not seem reasonable to state that a raw material does not have any value because it has not been processed yet. Clearly enough, without the raw material the end product or service cannot be produced. Just as without crude oil, there will be no petroleum products; without raw data, there will

[109] Facebook changed its slogan 'It's free and always will be' to 'It's quick and easy' between 6 and 7 August 2019. Moynihan and Asenjo (2019).

[110] Malcolm et al. (2017), p. 27.

[111] Malcolm et al. (2017), p. 27.

be no processed data. While a petroleum product undoubtedly has a higher value than crude oil, it is undisputed that this latter also has value.

The argument that raw data are valueless seems to be motivated by several factors. In the authors' opinion, three of such factors are especially relevant; namely undefined property rights, the outstanding amounts of data that are being generated every day and users' ignorance of the value of their data.

Albeit the mention of the term 'data ownership' in the literature as of 1981, to date no clear regulation relating to the problem of data ownership exists.[112] A *de facto* control of data is taking place due to existing technologies as well as a number of available legal tools, such as the right to data portability and the duty to obtain informed consent before personal data can be collected and used.[113] This *de facto* control translates into a *de facto* ownership of data. Although digital data are not formally recognised as property, they are treated as 'goods', in other words as if they are owned, in day-to-day commercial and economic life.[114] The current situation of *de facto* ownership functions on a first-come-first-served basis and seems to be governed by the law of the jungle, where the strongest party—i.e., the wealthiest, largest or most technologically advanced one—prevails.[115] The end result is an economy that treats data as a capital of data aggregators and not as a property or labour of users.[116] As a direct consequence of treating data as capital, a data market, where users commercialise the data that they own, fails to be created. Instead, data are collected by data aggregators, who become the only 'authority' to designate the approximate value of data. Not surprisingly, data aggregators often emphasise that raw data are valueless and that data become valuable exclusively after being processed.

The substantial amounts of data that are being generated and collected seem to constitute another factor behind the 'non-economic character' of raw data. While undisputedly being an extremely valuable raw material, their wide collection and storage cause an abundance of data, which, in turn, decreases their value. This phenomenon can be explained by the fact that value is not something that is inherent in things, no property in them.[117] As a matter of fact, value does not exist by itself;[118] it is something that we (as a society or a person) attribute to things.[119] When the demand for a particular item is much smaller than the available quantity, that item is likely to attain a non-economic character, which implies that it does not have value at

[112] Asswad and Marx Gómez (2021), p. 465. The article dates back to 2021, but to the extent known by the authors of this book such a regulation does still not exist.

[113] Janeček (2018), p. 1045.

[114] Swinnen (2020), p. 144.

[115] Swinnen (2020), p. 144.

[116] See Arrieta-Ibarra et al. (2018), pp. 39–40.

[117] Menger (2007), p. 116.

[118] Menger (2007), p. 121.

[119] Menger (2007), p. 116.

all or it has little value.[120] For example, if an inhabitant of a virgin forest has several hundred thousand trees at their disposal, while needing only 20 trees a year to satisfy their needs, a single tree will not have any value to them. If, however, the same individual has only 25 trees at their disposal, while still needing 20 trees on a yearly basis, even a single tree will have value to them.[121] In the same vein, in a developed country, a portion of drinking water has almost no value to an individual under ordinary circumstances, while a small piece of gold or diamond is likely to exhibit a rather high value. The reason for this discrepancy is the fact that individuals are not only in a position to satisfy their needs of drinking water fully, but also let large quantities of its escape unused, since they are unable to use up the whole available quantity;[122] while, on the other hand, diamond and gold are rare materials.[123]

The fact that users are not aware of the value created by their data seems to form the third main reason behind the consideration of raw data as valueless. Arrieta-Ibarra et al. argue that historical developments have made it attractive for siren servers to maintain data as capital equilibrium where users are unaware of the value created by their data on a daily basis. According to these contributors, if users were aware of the value of their data, they would likely demand compensation in a range of settings, which would dramatically reduce the share of the value that could be captured by the siren servers as profits.[124]

While the exact reasons for which raw data are considered to be valueless may not be established with certainty, one thing can: data aggregators are currently not paying for the 'raw material' that they utilise. The exact contours of the concept of 'value creation' employed in tax law are far from being clear, as it was already underlined by a number of commentators.[125] Whether it should be considered that users are creating value by generating raw data or not would be open to discussion and interpretation. The fact that the raw data obtained from users have value, on the other hand, is utterly clear and should not be subject to debate. This leads us to conclude that under the 'free data model', users are not being remunerated for

[120] Menger (2007), p. 116.

[121] Inspired from the example given by Menger (2007), p. 117.

[122] This example only aims to illustrate the value attached to a thing when there is an abundance of that particular thing. The intention is by no means to underestimate the overgrowing problem of water scarcity and crisis.

[123] Menger (2007), p. 140.

[124] Arrieta-Ibarra et al. (2018), p. 41.

[125] For a detailed explanation on this point, see Elliffe (2021), pp. 34–41.

generating data and enterprises collect their raw material for 'free'.[126] This constitutes a significant problem that should be remedied for.[127]

3.4.2 *Privacy Concerns*

From employing applications that can track online and offline behaviour to making use of the most mundane devices that are equipped with sensors; consumers are leaving a digital trace for almost every activity that they conduct. Such traces are then collected, stored and used by different data aggregators. This process gives rise to privacy concerns, predominantly for the three following reasons: (1) lack of legislative protection; (2) long period of storage; and (3) data breaches and misuses.

3.4.2.1 Lack of Legislative Protection

To address potential privacy concerns, some countries have enacted personal data protection legislations, the most notable example of which is the EU's GDPR.

It can be observed that the traditional approach on personal data protection has been to develop a core set of functional principles meant to adapt to various situations and technologies that are likely to emerge over time.[128] In its current state, personal data protection heavily relies on the requirement of 'informing individuals and obtaining their consent' when their personal data are being collected, used and/or disclosed to third parties. It is also typical for individuals to be granted the right to access their personal data held by the relevant entities, as well as to obtain the rectification of any inaccurate data. As summarised by Solove, "the goal of this bundle of rights is to provide people with control over their personal data, and through this control people can decide for themselves how to weigh the costs and

[126] In some cases, users enter into 'digital barters', i.e., they obtain 'free' digital services in exchange for providing their data. The fact that users may be compensated by means of digital services will not be mentioned each time the reference is made to 'free data model' or 'free collection of data' to avoid repetition and for the ease of reading. For more information on the digital barter, see Sect. 3.5.1 of this chapter.

[127] A similar observation has been made by several commentators. For example, Singh and Vipra underline that albeit the existence of a bias in hoping that the market itself would ensure economic fairness, in real-life such market logic does not seem to be working for the Uber driver or the merchant on Amazon.com [Singh and Vipra (2019), p. 53]. Along the same line, Lanier expresses that "the lion's share of wealth now flows to those who aggregate and route those offerings, rather than those who provide the 'raw materials'" [Lanier (2013), p. 9]. In a similar vein, Rees states that "if an aggregator is able with your permission, to make many useful inferences about your spending patterns and social behaviour so as to generate high advertising income focused on you, then the aggregator should find a way of rewarding you for the grant of your permission in the same way that the law rewards the grant of a licence of your property in other contexts" [Rees (2013), p. 221].

[128] Kerr and Szilagyi (2016), pp. 348–349.

benefits of the collection, use, or disclosure of their information".[129] The commentator refers to this approach of personal data protection as 'privacy self-management'.[130]

Regulating data protection by means of a core set of functional principles has revealed to be more or less successful to date, but it certainly has its limits. For instance, during the Covid-19 pandemic, a number of issues relating to personal data remained unclear, translating into an urgent need of further guidance.[131] In a similar vein, and with a more direct relation with the data market, substantial problems may also occur due to 'disruptive technologies', which undermine the functional principles of personal data protection. For example, 'informing individuals and obtaining their consent' method does not fit well in the context of ubiquitous computing. As a matter of fact, it is practically impossible to obtain meaningful consent for every ubiquitous computing transaction.

Regardless of their potential adaptation to different technologies and unexpected situations, personal data regulations in their current form may also simply choose not to, alternatively fail to, protect consumers data for several reasons. In this context, one of the most significant obstacles posed by personal data protection regulations is the definition of personal data itself. For instance, the data that have been rendered public by users themselves (e.g., via social media platforms) and the anonymous data (i.e., data that cannot be linked to an identified or identifiable person) are not considered as personal data. This may result in the non-applicability of personal data protection regulations—even when such regulations do exist and are efficient—to data aggregated by data brokers and by a variety of devices with sensors, without the knowledge of users. This lack of protection is clearly not a failure of the regulations concerned, but a policy choice.

Another important obstacle posed by personal data regulations, or rather their implementation, is the excessive reliance on 'consent', or as put by Solove 'privacy self-management'. Due to their market dominance, multi-sided platforms exercise a quasi-governmental authority that makes consumer consent to data collection meaningless in its essence.[132] The main reason behind this observation is that consumers may simply have no other option than to agree to the terms of service prepared by 'internet giants' in order to participate in the 'modern world'.[133] For example, some employers may consider that social media profiles are as crucial as résumés and deem consulting such profiles as one of the standards of the hiring process. From the perspective of such an employer, the fact that an employee does not have any social media profiles may be perceived as a red flag.[134] In a similar vein, if an online messaging service is being used by almost everyone as one of the primary means of

[129] Solove (2013), p. 1880.

[130] Solove (2013), p. 1880.

[131] See for example, European Data Protection Board (2020).

[132] Kim and Telman (2015), p. 728.

[133] Kim and Telman (2015), p. 769.

[134] Gillooly (2016).

communication, some users may feel compelled to use that service and may find themselves under the obligation to accept the terms of service prepared by the digital giant providing it.

Such pressure of participating in modern life may force users to go along with practices they otherwise disagree with. This tendency has been demonstrated on several occasions by a number of studies. For example, in a survey conducted in 2016 by Kesan et al., over 80% of the participants indicated that on some occasion, they had submitted information online when they wished that they did not have to do so.[135] Similarly, in a survey conducted in 2013 by Consumer Action, a non-profit consumer rights organisation, it was found that 69% of the respondents would not be willing to allow companies to track, collect, and share their data with their permission in exchange for a free product or service.[136] In yet another survey conducted in 2013 by Pew Research Center, it was observed that 86% of Americans have tried to hide their online activities by deleting cookies and encrypting e-mails and 55% of those surveyed have also 'taken steps to avoid observation by specific people, organisation or the government'.[137]

The 'information and consent' method does not constitute the most appropriate method for a variety of other reasons. First, consumers may not even realise that they consent to personal data collection and enter into legally binding agreements simply by clicking on an icon. In fact, consumers often indicate that they do not recall clicking on a link.[138] Second, terms of service may be formulated in a manner that leads consumers to think that the data collected are used only to their benefit, or at least in a manner that does not fully disclose the exact extent and the nature of the data mining.[139] Third, consumers are simply suffering from a general feeling of learned helplessness due to their inability to negotiate the terms of service prepared by multi-sided platforms dominating the market.[140] Finally, terms of service are often too long and hyperlinked to other web pages containing even more terms. It is simply impossible for consumers to be able to give an informed consent to all of the online contracts they enter.[141] As put by Kim and Telman, no human being has the cognitive capacity or the time to read through every digital contract they encounter.[142] In fact, according to a study conducted in 2008 by two researchers at Carnegie Mellon University, reading all the privacy policies on every website they visit would take a consumer approximately 25 full days on a yearly-basis, which translates into 76 work days a year.[143] On a national scale (from a US perspective) the total of hours

[135] Kesan et al. (2016), p. 267.

[136] Kim and Telman (2015), p. 739.

[137] Elvy (2017), p. 1379.

[138] Kim and Telman (2015), p. 732.

[139] Kim and Telman (2015), pp. 731 and 740.

[140] Kim and Telman (2015), p. 737; Kesan et al. (2016), p. 279.

[141] Kim and Telman (2015), p. 733.

[142] Kim and Telman (2015), p. 733.

[143] Madrigal (2012).

required to read the policies would amount to 53.8 billion, which would have translated into a national opportunity cost[144] of USD 781 billion, which is greater than the GDP of Florida, the fourth largest state economy in the US.[145]

Even if consumers could have overcome all these obstacles and made exceptionally well-informed and rational decisions, they could still not appropriately self-manage their privacy due to a number of structural problems.[146] The most notable issue in this regard is that many privacy harms are deriving from an aggregation of a variety of data over a period of time by different entities. In an environment where data are heavily collected, aggregated, processed and shared by a myriad of actors, it is virtually impossible for consumers to track all that 'data flow' and, thereby, to weigh properly the costs and benefits of revealing some information and/or permitting the use and transfer of such information.

3.4.2.2 Long Period of Storage

In the absence of legislation requiring entities to have data deletion protocols in place and/or a clause in the applicable contract accompanied with efficient control mechanisms, entities gathering data may be inclined to keep that data 'forever'. For example, some of the data brokers examined during the preparation of the Federal Trade Commission's 2014 Report, 'stored all data indefinitely, even if they are later updated'.[147]

While it is disputable to what extent storing data for exceptionally long periods may be beneficial for future business objectives, it certainly raises serious privacy concerns. Data aggregators constitute valuable targets for identity thieves and other similar actors, who are attracted to the collection of consumer information and/or profiles that could provide them with extensive information on a great number of persons.[148] On this point, it is important to underline that data breaches do not always occur due to a weakness in the data aggregator's security system; they can also materialise despite the implementation of all relevant measures. Data stored indefinitely translate into a risk with no time limits from a user point of view.

[144] Opportunity cost can be defined as the net benefit of time spent in the next best alternative activity [See, for example, Shaw (1992), p. 108].

[145] Madrigal (2012).

[146] Solove (2013), p. 1881.

[147] The US Federal Trade Commission (2014), p. vi.

[148] The US Federal Trade Commission (2014), p. vi.

3.4.2.3 Data Breaches and Misuses

Privacy concerns become a violation of privacy when a data breach or a data misuse occurs. In such cases, there exists a confirmed disclosure of data to third parties, and not merely a risk of exposure.

Data breach mainly refers to the nonintentional release of data to unauthorised third parties, generally due to the failure of entities to secure their databases. The concept of data breach can be construed in a larger manner depending on the context. For instance, article 4(12) of the GDPR defines 'personal data breach' as "a breach of security leading to the accidental or unlawful destruction, loss, alteration, unauthorised disclosure of, or access to, personal data transmitted, stored or otherwise processed". This definition encompasses three different types of breaches that follow the three well-known information security principles, namely confidentiality breach, integrity breach and availability breach.[149] Confidentiality breach occurs in case of an unauthorised or accidental disclosure of, or access to, personal data.[150] Integrity breach materialises in the event of an unauthorised or accidental alteration of personal data.[151] Finally, availability breach happens when there is an accidental or unauthorised loss of access to, or destruction of, personal data.[152]

For the purposes of the data market, the most relevant breach from a privacy point of view is 'confidentiality breach' as it is being referred to in the context of GDPR, or simply data breach in its broad sense (i.e., nonintentional release of data to unauthorised third parties). Multi-sided platforms, data brokers and other data aggregators may have sensitive information on users, which can cause serious privacy concerns in case of a nonintentional release. Accordingly, in this book the term 'data breach' refers to nonintentional release of data to unauthorised third parties.

There have been several large-scale data breaches in the last decade. Are amongst the most notable examples: the Yahoo data breaches, which reportedly compromised one billion user accounts in 2013 and 500 million user accounts in 2014;[153] the Equifax data breach, which reportedly compromised sensitive information for 143 million American consumers—including Social Security numbers and driver's license numbers[154]—and the Target data breach, which reportedly compromised names, credit or debit card numbers and three-digit security codes for 40 million consumers.[155]

[149] European Commission (2018), p. 7.

[150] European Commission (2018), p. 7.

[151] European Commission (2018), p. 7.

[152] European Commission (2018), p. 7.

[153] Goel and Perlroth (2016).

[154] Bernard et al. (2017).

[155] Satariano and Confessore (2018).

Data misuse, on the other hand, refers to the phenomenon whereby the use of the data by the data aggregator amounts to a legal and/or an ethical problem. Data misuse is not limited to but comprises an intentional release of data to third parties by the data aggregator. In this contribution, the term 'data misuse' refers to such intentional release of data. The most prominent example of data misuse is, beyond any doubt, 'the Cambridge Analytica scandal'. The now-defunct political consulting firm Cambridge Analytica, the services of which were employed during Brexit and the Trump campaign, had obtained and exploited Facebook data from as many as 87 million users around the globe. It was discovered that the firm created a 'psychographic profile' of voters ahead of the 2016 US election on the basis of the data mined from the Facebook accounts of users, and employed those profiles to engage in behavioural microtargeting to deliver specific pro-Trump messages that resonated with specific voters for highly personal reasons.[156]

Prima facie, the effects of a data breach or misuse seem to be limited to users whose data were disclosed to or accessed by unauthorised third parties. As also demonstrated by the Cambridge Analytica scandal, such a narrow view is far from reflecting the reality. The harms deriving from data breaches and misuses are often far greater than the simple sum of private injuries to the individuals whose data are obtained by third parties. On this basis, Ben-Shahar argues that "data pollution is to our century what industrial pollution was to the last one".[157] As discussed in detail by Ben-Shahar, data breach and misuse cases give rise to repercussions affecting the entire ecosystem—i.e., the society as a whole—and not merely individuals whose data are emitted or misused.[158] The contributor regroups such repercussions into three broad categories: effects on public interests, effects on other individuals and precaution and insurance externalities.

Effects on public interests can materialise as a primary threat to the public good (this concept refers to the misuse of data producing an impact on the core constituents of a country, such as interfering with the elections and sharing location information that can compromise national security), harmful discrimination and the rapid dissemination of manipulative and polarising news and beliefs.[159]

Effects on other individuals are produced when users provide specific information about other individuals to data aggregators. For instance, a Gmail user is allowing Google to collect and use personal data of other individuals (who are not Gmail users), given the fact that the company allegedly scans the texts of Gmail messages sent and received by its users.[160]

[156] Galloway (2017), p. 105.

[157] Ben-Shahar (2019), p. 106.

[158] Ben-Shahar (2019), p. 112.

[159] Ben-Shahar (2019), pp. 112–115.

[160] Ben-Shahar (2019), pp. 115–116.

Precaution and insurance externalities simply refer to costs incurred to prevent some of data's external effects.[161] The burden of such costs is, in fact, not being ultimately borne by private parties (i.e., individuals and entities) directly suffering from a data breach or misuse. These costs constitute 'public expenditures', not in a technical sense but in the sense that they are borne by all consumers. For instance, when a security breach occurs and sensitive personal data are released, individuals whose data have been released could suffer significant private harm in the form of identity theft and financial fraud. However, at least in some countries, individuals are largely insured against these losses. As a result, consumers whose data are stolen may end up suffering from only a small fraction of the economic costs of a data breach or misuse. The full economic burden of the breach will, in turn, be borne by all consumers, due to its translation into higher fees and prices for other products and services available on the market.[162] In the same vein, the economic impact of data breaches may cause companies affected by such a breach to increase the price of their services or products. In the research conducted by Ponemon Institute and published by IBM Security, 60% of the studied organisations reported proceeding to such an increase in order to make up for the costs incurred by a data breach, the average cost of which amounted to USD 4.35 million in 2022.[163]

3.4.3 Jobs Are Disappearing

Whether AI is reducing, and will further reduce, the number of available jobs on the market and will provoke an employment crisis constitutes one of the most popular debate subjects of the last decade. The discussion on the issue seems to have been fueled with the recent development of conversational and collaborative models of AI, using large language models, such as the widely acclaimed ChatGPT.[164] While these conversational models are not necessarily revolutionary from a technological perspective, they are easily accessible to all, able to answer questions in natural language, and apt to be employed for a great variety of tasks.[165]

Not surprisingly, a number of studies were conducted on the 'disappearing jobs phenomenon' and the results they reached were not very optimistic. One of the most renowned studies in the field is the one conducted by Frey and Osborne of the University of Oxford, published in 2013, according to which as many as 47% of the

[161] Ben-Shahar (2019), p. 116.

[162] Ben-Shahar (2019), p. 118.

[163] IBM (2022), p. 5.

[164] ChatGPT, more precisely 'Chat Generative Pre-Trained Transformer', developed by the firm OpenAI, has been introduced in the form of a system of generative artificial intelligence, based on large language models. The key abilities of ChatGPT are to understand and efficiently interact in a human language as well as to learn through experience (the so-called 'deep learning').

[165] For more information on large language models, see, for example, French Government (2023), pp. 6–9.

total US employment is in the high-risk category, meaning that the associated occupations are potentially automatable over some unspecified number of years.[166] Another study conducted in 2017 estimates that around 18% of the existing United Kingdom (UK) jobs could face a high probability (over 70%) of automation over the next 10 years, rising around just under 30% after 20 years.[167]

These findings can be illustrated by real-life examples from the companies active in the data market. According to Arrieta-Ibarra et al., the market capitalisation and value-added by firms like Facebook, Google and Microsoft are similar to, or even greater than, a firm like Walmart, yet they employ one to two orders of magnitude fewer workers and the publicly available statistics of these companies suggest that their labour income shares are a small fraction of the traditional average of 60–70%.[168] For instance, Facebook pays only about 1% of its value each year to its workers, while Walmart pays out 40% of its value in wages.[169] In a similar vein, Kodak employed more than 140,000 people and was worth USD 28 billion at the height of its power, while Instagram employed only 13 people when it was sold to Facebook for a billion dollars in 2012.[170]

While a certain percentage of the available jobs, or certain portions of them, will certainly disappear due to automation, a variety of new jobs will be created. Some of these new jobs are expected to be in areas directly linked to AI and related technologies (such as data scientists and robotic engineers).[171] However, it is foreseen that most of the additional employment will not be in 'high tech' areas, but rather in provision of hard-to-automate services (such as health and personal care).[172] The persisting, and the essential, question is, what will be the overall net effect on the employment market. The answer to this question remains quite unclear to date.

As explained by Acemoglu and Restrepo, the exact impact of automation on labour market and productivity depends on several factors, the outcome of which cannot be predicted with certainty. Clearly, AI and robotics create a powerful displacement effect. This effect may, however, be pushed against by other factors, mainly productivity effect, capital accumulation, deepening of automation and reinstatement effect. The overall balance of these conflicting forces will determine the ultimate outcome of the impact of automation on labour market.[173]

Automation transforms the production process to permit more *tasks* to be performed by machines.[174] To properly understand this phenomenon, it is necessary

[166] Frey and Osborne (2013), p. 38.
[167] PwC (2021), p. 6.
[168] Arrieta-Ibarra et al. (2018), p. 38.
[169] Posner and Weyl (2019), p. 71.
[170] Lanier (2013), p. 18.
[171] PwC (2021), p. 6.
[172] PwC (2021), p. 6.
[173] Acemoglu and Restrepo (2018), pp. 1–3.
[174] Acemoglu and Restrepo (2018), p. 5.

to distinguish 'jobs' from 'tasks'. Jobs, especially knowledge jobs (as opposed to manual jobs), consist of a collection of smaller tasks. Automation may replace only a portion of a job—i.e., some of the tasks composing the job—while not replacing the job as a whole.

In the eventuality where capital is sufficiently cheap or sufficiently productive at the margin, the automation will lead to the substitution of capital for labour in tasks that were subject to automation. Such a substitution results in a *displacement* of workers from the tasks that are being automated.[175] While such a displacement may be limited merely to a number of tasks, leading thereby to a partial replacement, it still produces a negative effect on the workers concerned. As a matter of fact, in contrast to the standard approach based on factor-augmenting technological changes, a task-based approach immediately opens the way to productivity-enhancing technological developments that simultaneously reduce wages and employment.[176] Despite being often ignored, technological improvements that increase productivity also risk to reduce the wage of *all workers*.[177]

The displacement effect that can currently be observed is quite powerful, visible and disturbing. As per an estimate made by Galloway in 2017, 'the four' (i.e., Amazon.com, Apple, Facebook and Google) employ approximately 780,000 employees, which roughly corresponds to the population of Charlotte, the most populous city in the US State of North Carolina. The combined value of the four companies' public shares of stocks amounts to approximately USD 2.7 trillion, which is almost the equivalent of the GDP of France, a developed nation of 67 million citizens. By comparison, Disney, commands a market capitalisation of USD 169 billion and yet employs 195,000 persons and Intel has a market capitalisation of USD 219 billion and has 107,000 employees.[178] In the face of these numbers, Galloway points out that "We have a perception of these large companies that they must be creating a lot of jobs, but in fact they have a small number of high paying jobs and everybody else is fighting over the scraps. America is on pace to be home to 3 million lords and 350 million serfs".[179]

The presence of the *displacement effect* does not automatically translate into a reduction of the number of available jobs on the market. The displacement effect can be *counterbalanced* by other factors, namely productivity effect, capital accumulation, deepening of automation and reinstatement effect.

- *Productivity effect*: As automation reduces the cost of producing a subset of tasks, the prices of the goods and services whose production processes are being automated may also be reduced. This would, in turn, make households effectively richer and increase the demand for all goods and services. Such an expansion of

[175] Acemoglu and Restrepo (2018), pp. 5–6.
[176] Acemoglu and Restrepo (2018), p. 6.
[177] Acemoglu and Restrepo (2018), p. 6.
[178] Galloway (2017), pp. 252–253.
[179] Galloway (2017), p. 254.

the economy will have the effect of increasing the demand for labour in non-automated tasks. The productivity effect may not be able to push against the displacement effect, if automation technologies that are adopted are just productive enough to cause displacement, but not sufficiently productive to bring about powerful productivity effects.[180] This may reveal to be a serious problem in the field of AI and robotics, given the fact that, for the first time in history, innovation does not seem to be targeted at complementing labour by making it more productive, but rather aims to replace human power.[181] This tendency may easily be observed by taking a closer look to the investments made by big tech companies.[182]

From a data market perspective, it can be stated that its current functioning (which can be referred to as 'the free data model') has made productivity-related data much less accessible than consumption-oriented data. AI may have remarkable potential, but in practice, it is predominantly fuelling targeted advertisement rather than improvements in productivity. In other words, a purely free data economy acts as a drag on productivity growth.[183]

- *Capital accumulation*: Automation is characterised by an increase in the capital intensity of production. The high demand for capital is expected to trigger further accumulation of capital, which then may raise the demand for labour.[184]
- *Deepening of automation*: Technological improvements may also increase the productivity of capital in tasks that have already been automated. Since labour was already replaced by capital in those tasks, an additional displacement is not created. This deepening of automation gives rise to a productivity effect, which then may raise labour demand.[185]
- *Reinstatement effect*: The creation of new labour-intensive tasks, which engender a reinstatement effect, has the potential of forming the most powerful force balancing the growth process in the face of rapid automation.[186] As a matter of fact, this effect acts as a direct counterforce to displacement effect.

From a historical perspective, it can be observed that ultimately the reinstatement effect was powerful enough to counterbalance the displacement effect. What must not be overlooked, however, is the *painful nature of the adjustment*. For instance, the introduction of new technologies during the British Industrial Revolution ultimately led to rising labour demand and wages, but only after an 80-year period of stagnant wages, expanding poverty and harsh living conditions.[187] This cumbersome process

[180] Acemoglu and Restrepo (2018), pp. 6–7.

[181] Huettinger and Boyd (2020), p. 41.

[182] For the example of Amazon.com, see Galloway (2017), pp. 31–51.

[183] Arrieta-Ibarra et al. (2018), p. 39.

[184] Acemoglu and Restrepo (2018), pp. 7–8.

[185] Acemoglu and Restrepo (2018), p. 8.

[186] Acemoglu and Restrepo (2018), pp. 9–10.

[187] Acemoglu and Restrepo (2018), pp. 12–13.

of adjustment can be explained by several factors, among which the mismatch between the skills of the labour force and the skills required by new tasks. To continue with the same example, the wages started growing in the nineteenth-century British economy only after mass schooling and other investments in human capital expanded the skills of the workforce.[188] Luddites, who seem to be perceived as a crowd who failed to understand the opportunities that were provided by automation, were actually not smashing power looms during the first industrial revolution out of sheer irrational fear. They assumed that their textile jobs were at risk and their assumption was confirmed; the machines did take over their jobs.[189] Even though new jobs were created following the first industrial revolution, when and under which circumstances Luddites could find other jobs is beyond the authors' knowledge, since it is not mentioned as frequently as their protest, if it is mentioned at all.

Some contributors express their concerns about the fact that this industrial revolution diverges from the previous ones for a variety of reasons and it may not result in the creation of a sufficient number of new jobs.[190] Some other commentators opt for the opposite view and argue that new jobs will be created and there are no serious reasons to be overly pessimistic.[191] Independent of the 'winning argument', which can only be established with certainty in due time, one thing is clear: even if a sufficient number of new jobs, or alternatively tasks, are created in time and the automation does not ultimately provoke an unemployment crisis, the adjustment period will be slow and painful, as it was during the previous industrial revolutions. As a matter of fact, creation of new jobs may even be slower in this industrial revolution. One of the main reasons for this prediction is the almost singular focus in the corporate sector and research community on AI, applications of deep learning, and other big data methods to automate various tasks. Too much attention is being devoted to automation, which reaches proportions that are larger than socially desirable. Such excessive automation produces a negative impact on other productivity-enhancing technological breakthroughs and thereby productivity.[192]

For the reinstatement effect to counteract the displacement effect in an efficient manner, the newly created jobs (or alternatively tasks) should not result in dire working conditions and lower wages when compared to 'traditional' jobs. With regards to this issue, Gray and Suri project that at the current rate of growth combined with current trends in the growth of contract staffing and temporary agency services, 60% of today's global employment is likely to be converted into some form of microwork by 2055.[193] While microwork clearly amounts to 'new tasks', the working conditions and wages of microworkers are far from even

[188] Acemoglu and Restrepo (2018), p. 13.

[189] Malcolm et al. (2017), p. 14.

[190] See, for example, Galloway (2017), p. 51.

[191] Malcolm et al. (2017), pp. 34–38.

[192] Acemoglu and Restrepo (2018), pp. 14–15.

[193] Gray and Suri (2019), p. 179.

remotely matching the working conditions and wages of a typical employee. Stating that such new tasks are offering a viable post-automation employment option would be an overreaching argument.

As demonstrated by research conducted in the field, scientific analysis and real-life examples, a non-negligible number of jobs *are currently disappearing* due to automation. This displacement effect can be counterbalanced in the future. It is not possible to predict with certainty when the displacement effect will be counteracted—if ever—and to what extent. What is certain, however, is that an adjustment process will be needed and such process is likely to be long and painful.

3.4.4 Operational Local Scale Can Be Reached Without Local Mass

As it is common knowledge, globalisation allowed businesses to locate various parts of their production process across different countries and at the same time access a greater number of customers around the globe.[194] This trend intensified with digitalisation. By means of using remote technologies, many digitalised enterprises can effectively be heavily involved in the economic life of a great number of jurisdictions without any, or any significant, physical presence.[195] The enterprises concerned can thus achieve operational local scale without necessitating any local mass (cross-jurisdictional scale without mass).[196]

The non-relevance of the physical presence in the context of the digital economy can be illustrated by means of a multi-sided platform (see Fig. 3.3). For instance, an Irish platform may provide 'free' digital services to French consumers, collect their data in exchange, use that data to provide targeted advertisement services to French companies by displaying targeted advertisement to French consumers. While the great majority of the economic activity is intrinsically linked to France, the physical place of business is situated in Ireland. Parts of the Irish platform's economic activity linked to France may derive from different jurisdictions as well. For example, German companies may display their advertisements to French consumers by means of the services provided by the Irish platform.

On this basis, it can be concluded that the need to physically perform the services in the country where the customers and/or the consumers are residing is systematically eliminated in the context of the digital economy with regard to a non-negligible number of sectors. The removal of the physical presence renders the assignment of the relevant activities to different jurisdictions more and more

[194] OECD (2018), § 131.

[195] OECD (2018), § 132.

[196] For further explanations on cross-jurisdictional scale without mass, see notably OECD (2018), § 33.

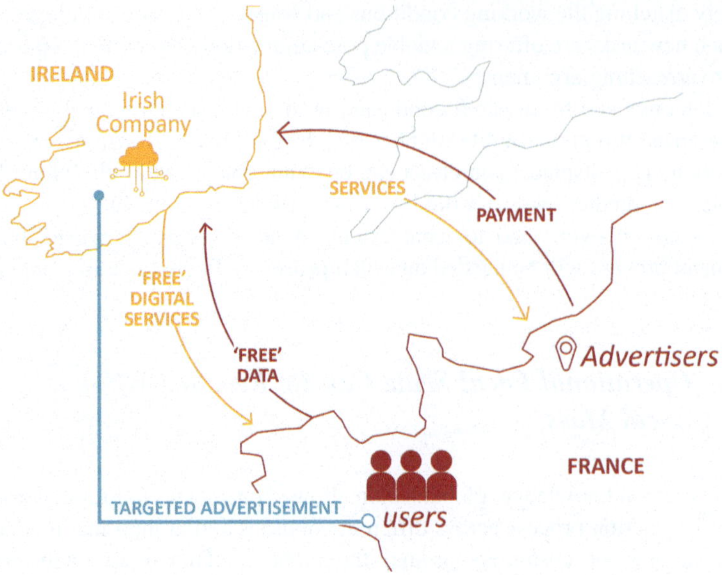

Fig. 3.3 A typical example of activities conducted by 'digital' multi-sided platforms

complex, and that complexity gives rise, in turn, to the necessity to adjust the current tax law rules to the new economic reality.[197]

3.5 Identifying the Most Typical Uses of Data

Conducting a proper tax law analysis of the data market necessitates a clear identification of the most typical transactions taking place between the actors active in the market concerned as well as of the other main uses of data. Different operations involving data can be regrouped into four steps (see Fig. 3.4):

- *Step one* covers operations related to data generation by users and data collection by data aggregators. This step occurs in generation and collection phases of the data's life cycle.
- *Step two* consists of microwork, which takes place in processing and analytics phases of the data's life cycle.

[197]For more information on this issue, see Chaps. 5 and 8.

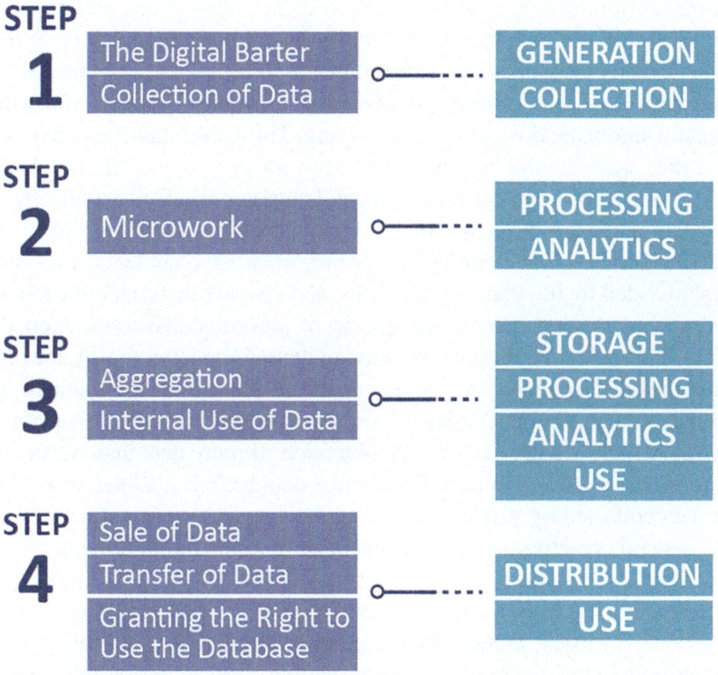

Fig. 3.4 Most typical operations involving data regrouped into four steps

- *Step three* involves analysis and use of data by enterprises for internal purposes. Such operations relate to storage, processing, analytics and use phases of the data's life cycle.
- *Step four* encompasses, among others, sale of data, transfer of data and granting the right to use a database. These operations relate to distribution and use phases of the data's life cycle.

This section aims to provide general information relating to these four steps and to identify the main issues arising in each step, to serve as a basis for the tax law analysis that is conducted in the following chapters. Examining all potential uses of data being out of the scope of this contribution, are only mentioned the most typical uses that are relevant from a tax law point of view.

3.5.1 Step One

Step one regroups data generation by users and data collection by data aggregators. This generation/collection mainly occurs in four different manners. First, users may share their data in the course of a transaction they are entering into with a data aggregator. For instance, a client purchasing a car from a car manufacturer shares several pieces of information with this latter, such as their name, address and phone number. Second, user data may continuously be collected by a device or an equipment. As already explained, this phenomenon is a direct result of the IoT. To continue with the same example, a consumer's car may collect information and submit them automatically to a group company. Third, user data may be collected by different data aggregators, beyond users' knowledge and without any direct commercial, or other, relations between users and aggregators. This is typically the case of the data aggregated by data brokers. Fourth, users may exchange their data for 'free' digital services. For example, a user may open a social media account, use the services provided by the platform for 'free' and provide their data in exchange.

Prima facie, sharing data in the course of a commercial transaction does not constitute, at least not predominantly, a use of data. As a matter of fact, the purpose is not to use the data *per se*, but to make use of the necessary data to be able to complete the transaction that is being conducted, in our example a car sale. This main purpose does, however, clearly not exclude the obtention of more data than necessary, their aggregation, their use and even their sale to a data broker at a later stage.

The data collected by various devices and equipment are directly destined to be used for several objectives. The data generated by users through the devices can be employed for internal purposes by the company or sold to data brokers. Internal purposes encompass, among others, understanding users' preferences, providing better services to users, enhancing the product through the feedback that is constantly and automatically received and providing services to the other side of the platform. A car manufacturer may, for instance, observe malfunctions and potential improvements through the information transmitted by its cars.

In a similar vein, data brokers gather data with the direct purpose of using them for their commercial operations. A data broker may, for instance, infer that a user is about to purchase a sports car on the basis of other information it could gather on that same user (websites visited, recent locations, Instagram posts etc.), label the user as an 'imminent sports car purchaser' and sell that labelled information to interested parties (car manufacturers and/or car dealers).

Exchange of data for 'free' digital services constitutes not just a typical but also the most debated use of data. As indicated by Hansell in The New York Times as early as 1999, "to get freebies, customers often must also disclose a dossier's worth of personal information for use by marketers and put up with minimal customer service".[198] This exchange has been dubbed as a barter by Litman in 2000,[199] who states: "Most obviously, there is the *implicit barter system* that characterizes many consumer transactions, and is exemplified by commercial World Wide Web sites. Information, goods, and services are free and freely available in return for a little volunteered demographic information, a dollop of attention and the ability to record, slide, dice, correlate and sell the clickstream data generated by your forays onto the

[198] Hansell (1999).

[199] The article written by Litman contains the earliest mention of 'digital barter' that could have been identified by the authors [see Litman (2000)].

Internet".[200] The barter can indeed be qualified as *implicit*, since users are perceiving much of the free digital services they use as entertainment, without clearly—and constantly—realising the true nature of the economic relationship and they are neither aware of the exact productive value of their data nor the role they play in enabling ML.[201]

In the authors' opinion, the term that best describes the barter consisting of free data generated by users for the purposes of being collected by the enterprises in exchange for free digital services provided by the enterprises concerned is '*the digital barter*'.[202] Reference is also frequently made to data-as-payment to describe the same phenomenon.[203] In the 'digital barter', users are not paid for their data contributions to digital services nor do they pay directly for the value they derive from these services.[204] The digital services that may be perceived as being free by users, are, thus, not 'free'. As a matter of fact, users are engaging in multiple 'micro-barter' transactions in which data about their activities and preferences are being exchanged for digital services.[205]

The digital barter constitutes an *imperfect barter* transaction in which data are swapped for a digital service. While it is possible that the value of the service provided is higher than the value of the data obtained, it is also possible that the value of the data exceeds the value of the service provided.[206] It is important to note that under the freemium model (i.e., the 'free' and 'premium' model under which the company offers some additional services—the so-called 'upgrades'—on top of free services in exchange of subscription fees); users are still exchanging their data. In other words, consumers are paying with their data on top of the subscription fees.

Data do not form the only asset users are exchanging in the context of the digital barter. Data generation requires users not only to share their habits, personal information, emotions and other information, but also to sacrifice their most important 'possession': time. In 2015, Facebook's 1.5 billion users spent on average 50 min per day on the platform.[207] As a matter of fact, according to a study conducted in the US in 2018, the hours spent by an individual on social media amount to approximately 382 per year.[208] Users' time is actually an important, and yet often forgotten, component of the digital barter. "Companies would rather have consumers' time than their money", for several reasons.[209] First capturing users' attention may render the multi-sided platform more attractive for the customers of the other side. Second, the more active a user is, on a platform, device or application,

[200] Litman (2000), p. 1299. Emphasis added by the authors.

[201] See, for example, Arrieta-Ibarra et al. (2018), p. 41; and Posner and Weyl (2019), pp. 78–79.

[202] The term is used by Aslam and Shah (2020), p. 9.

[203] Elvy (2017), p. 1383.

[204] Arrieta-Ibarra et al. (2018), p. 38.

[205] Aslam and Shah (2020), p. 9.

[206] Elvy (2017), p. 1385.

[207] Posner and Weyl (2019), p. 78.

[208] Leack (2018).

[209] Hansell (1999).

the more they can generate data that can potentially be useful for data aggregators. This need for users' time and attention also emphasises the relevance of maintaining a predominantly 'entertaining' or 'useful' image.

Users' 'lost hours' give rise to an *opportunity cost*, which can be described as the net benefit of time spent on the next best alternative activity.[210] Calculating the opportunity cost is challenging. The cost may vary on the basis of different factors, such as the employment status of the individual, time of the year and the recreational or commercial nature of the activity.[211] While estimating an approximate opportunity cost is burdensome, it is clear that users may engage in much more relevant activities instead of spending time on a variety of online platforms. For instance, an average US resident may learn to drive by using only 18% of the time (i.e., 67 h) they spend on social media.[212]

While estimating the full impact of the digital barter on users, *consumer surplus* should also be taken into account. It is important to emphasise that consumer surplus refers to the difference between the consumers' willingness to pay for a good or a service and the amount they actually pay. For example, if a person is willing to pay up to USD 15 for a book but only had to pay USD 10, then that person would gain USD 5 of consumer surplus from that transaction.[213] In the context of 'free' digital services, establishing consumer surplus is a difficult task, given the fact that users are not paying with currency but with their data. Not taking data into consideration and deem that users are obtaining the digital services for free would result in an erroneous appreciation of the consumer surplus. The amounts that users are willing to pay for digital services may, however, be measured to a certain extent. This is precisely what has been effectuated by Brynjolfsson et al., who conducted a series of choice experiments to establish the value attributed by consumers to digital services. As per the research, from a user perspective, the median value for yearly access amounted to USD 17,530 for search engines, USD 8414 for e-mail, USD 3648 for digital maps, USD 1173 for videos and USD 842 for e-commerce in 2017.[214] Needless to remind, these amounts correspond to users' willingness to pay. Negative externalities that may produce a non-desirable impact on other people's well-being, adverse effects on the users' own mental or physical health, which users do not fully appreciate, and other similar factors were probably not taken into consideration by the users.[215]

To summarise, the digital barter consists of exchanging users' data and time with digital services provided by companies. Before concluding, it is essential to mention that the barter qualification may not be deemed appropriate by some commentators. It has already been argued that the 'exchange' between users and entities providing 'free' digital services does not constitute a barter, but simply corresponds to the

---

[210] Shaw (1992), p. 108.

[211] For a detailed explanation, see, for example, Shaw (1992).

[212] Leack (2018).

[213] Brynjolfsson et al. (2019), p. 7251.

[214] Brynjolfsson et al. (2019), p. 7252.

[215] Brynjolfsson et al. (2019), pp. 7253–7254.

functioning of the multi-sided platforms. This argument is also supported by data's non-rival nature.[216] In our view, the existence of a 'barter' is quite straightforward. Barter is the exchange of one good or service for another.[217] Users are exchanging their data and time for digital services. As also implied by the well-known 'adage', "if you are not paying for the product, you are the product", users cannot refuse to do so; they are under the obligation to exchange their data and time for the services concerned. As a matter of fact, companies are not rendering any digital services without collecting users' data and capturing their attention. Hence, the clear existence of a barter.[218]

The main issue arising in this first step is the lack of proper compensation provided to users for the data they generate. This issue is further discussed in Chaps. 5 and 6.

Qualification of the exchange of users' data with digital services as a barter gives inevitably rise to the existence of taxable transactions. As explained in detail, in the context of the digital barter, users are selling their data to entities and the latter are selling their digital services to the former. From a theoretical perspective, users should report their income and pay an income tax on the sale of their data and entities should report their income and pay a corporate income tax on the sale of access to their digital products.[219] As described in Chap. 10, the provision of 'free' digital services also constitutes a taxable transaction for VAT purposes, under a purely theoretical analysis. Whether such a taxation would be justified from a policy standpoint and a practical point of view, despite being plausible from a purely theoretical perspective, is subject to discussion. In the authors' opinion, which is described in Chap. 6, the proper approach would be to put an end to the 'free' collection of data. In this approach, the digital barter would also cease to exist, along with all the theoretical tax law problems it may raise.

3.5.2 Step Two

The main issues relating to microwork, such as working conditions and lack of legal protection, are mostly related to employment law and, thereby, the most appropriate remedies to the microwork phenomenon should be discussed from an employment law point of view, which is beyond the scope of this contribution.

---

[216] See, for example, Omaar (2021) and McQuinn (2018).

[217] See, for example, Mankiw (2017), p. 604.

[218] For the same view see, Litman (2000), p. 1299; Molina-Jimenez et al. (2020), p. 6; and Goodson and MarketShare (2012). Although it was not incorporated into the final reading, the Digital Content Directive (COM (2015) 634) proposed by the European Commission indicated that users share their data in exchange for access to digital content (art. 2). For further information, see, Duch-Brown et al. (2017), p. 17.

[219] See also Thimmesch (2016), pp. 15–20.

In the context of this book, microwork constitutes one of the side effects of excessive automation on labour market. Thereupon, *the main issue arising in this second step* from a tax policy perspective is the ever-increasing risk of unemployment and the necessity for governments to support their citizens in different manners until the displacement effect can be counterbalanced by other factors. This issue is elaborated in Chap. 7.

3.5.3 Step Three

Data that are collected by entities, especially by multi-sided platforms, may be processed in order to be used for different internal operations. The analysed user data are typically used by multi-sided platforms to provide services to the customers of the other side(s) of the platform (such as targeted advertisements) and to train the AI systems through ML.[220] In fact, AI runs on ML systems that analyse substantial amounts of user-generated data. Specific information and production services are designed by programming the interaction between data generators and machines. In other words, most of the difficult work is not deriving from profound algorithmic designs; instead, it involves tweaking the existing models to fit the relevant data and deliver the desired product or service.[221]

Users' data also constitute an efficient tool allowing to receive constant feedback from consumers, on the basis of which companies may proceed to the relevant adjustments of their products and/or services.

The use of data for internal purposes of a data aggregator becomes problematic from a tax law perspective primarily in the context of cross-border operations. With regards to the data collected and processed by multi-sided platforms, *the key issue* that comes to mind, and that has been widely discussed in the last decade, is the case of a multi-sided platform that collects information from the user base of a country and renders services in relation to that user base without paying an adequate amount of corporate income tax in the country concerned. This issue has been examined in detail by a myriad of commentators, private organisations and other institutions in the context of the 'taxation of digital economy', which has been, beyond any doubt, the most prominent discussion subject of international tax law in the last decade. With regard to collection and use of data for the purposes of performance analysis, *the most controversial issue* from a tax law perspective arises when user data are being collected by an entity in one jurisdiction and later on transferred to an affiliated entity situated in another jurisdiction.

From an international tax law point of view, the underlying problems relating to these two issues differ from each other to a great extent, albeit being linked to each other to a certain degree. In the context of a multi-sided platform making use of

---

[220] Posner and Weyl (2019), p. 74.

[221] Posner and Weyl (2019), p. 71.

consumer data in a jurisdiction where it is not liable to pay corporate income tax, the main obstacle for taxation is the *lack of a nexus*. On the other hand, in the case of a cross-border intragroup transfer of data, the main concern is to determine whether the transfer is being effectuated in conformity with the *transfer pricing guidelines*.

Given the significant degree of divergence between their underlying rationale, these two main internal uses of data are analysed separately. The lack of nexus is examined in Chap. 8 and the cross-border intragroup data transfers are examined in Chap. 9, along with other transfers of data.

3.5.4 Step Four

Data collected by entities may be sold to third parties or transferred to group companies in the form of raw data or processed data. For instance, a car manufacturer can sell raw data to a data broker. Similarly, a car manufacturer's subsidiary can collect data from the cars that are being used in the country where it is situated and transfer that raw data to a group company located in another country, which aggregates and analyses data coming from subsidiaries of different countries and submits this processed data to the manufacturer's headquarters situated yet in another country. Data aggregators may also create databases and grant third parties the right to use the databases concerned.

The main issues arising in this fourth step from a tax law perspective are the characterisation of income deriving from sale of data and from granting the right to use a database to a third party as well as the valuation of data transfers within group companies (transfer pricing). These issues are examined in Chap. 9.

References

Acemoglu D, Restrepo P (2018) Artificial intelligence, automation and work. NBER working paper series no. 24196, National Bureau of Economic Research

Arrieta-Ibarra I, Goff L, Jiménez-Hernández D et al (2018) Should we treat data as labor? Moving beyond "free". AEA Papers Proc 108:38–42

Aslam A, Shah A (2020) Tec(h)tonic shifts: taxing the "digital economy". IMF staff discussion note, IMF

Asswad J, Marx Gómez J (2021) Data ownership: a survey. Information 12(11):465

Ben-Shahar O (2019) Data pollution. J Leg Anal 11:104–159

Bernard TS, Hsu T, Perlroth N et al (2017) Equifax says cyberattack may have affected 143 million in the U.S. The New York Times. https://www.nytimes.com/2017/09/07/business/equifax-cyberattack.html. Accessed 07 July 2023

Brynjolfsson E, McAfee A (2014) The second machine age, 1st edn. W. W. Norton & Company

Brynjolfsson E, Oh JH (2012) The attention economy: measuring the value of free digital services on the Internet. In: 33rd International conference on information systems, Orlando

Brynjolfsson E, Collis A, Eggers F (2019) Using massive online choice experiments to measure changes in well-being. PNAS 116(15):7250–7255

Caldari K, Masini F (2011) Pigouvian versus Marshallian tax: market failure, public intervention and the problem of externalities. Eur J Hist Econ Thought 18(5):715–732
Collin P, Colin N (2013) Task force on taxation of the digital economy, Report to the Minister for the Economy and Finance, the Minister for Industrial Recovery, the Minister Delegate for the Budget and the Minister Delegate for Small and Medium-Sized Enterprises, Innovation and the Digital Economy
Cross T (2017) Beyond Moore's law. In: Franklin D (ed) Megatech technology in 2050, 1st edn. The Economist Profile Books, pp 55–63
Dube A, Jacobs J, Naidu S et al (2018) Monopsony in online labor markets. NBER working paper series no. 24416, National Bureau of Economic Research
Duch-Brown N, Martens B, Mueller-Langer F (2017) The economics of ownership, access and trade in digital data. JRC digital economy working paper 2017-01, European Commission Joint Research Centre, European Union
Elliffe C (2021) Taxing the digital economy: theory, policy and practice, 1st edn. Cambridge University Press Cambridge Tax Law Series
Elvy S (2017) Paying for privacy and the personal data economy. Columbia Law Review 117(6): 1369–1454
European Commission (2018) Article 29 Data Protection Working Party, guidelines on personal data breach notification under regulation 2016/679, adopted on 03 October 2017. https://ec.europa.eu/newsroom/document.cfm?doc_id=47741. Accessed 07 July 2023
European Data Protection Board (2020) Guidelines 04/2020 on the use of location data tracing tools in the context of the COVID outbreak. https://edpb.europa.eu/sites/default/files/files/file1/edpb_guidelines_20200420_contact_tracing_covid_with_annex_en.pdf. Acesssed 07 July 2023
French Government (2023) ChatGPT ou la percée des modèles d'IA. Éclairage sur…# 6. April 2023. https://www.peren.gouv.fr/rapports/2023 04 06_Eclairage%20sur_CHATGPT_FR.pdf. Accessed 07 July 2023
Frey CB, Osborne MA (2013) The future of employment: how susceptible are jobs to computerisation. https://www.oxfordmartin.ox.ac.uk/downloads/academic/The_Future_of_Employment.pdf. Accessed 07 July 2023
Galloway S (2017) The four: the hidden DNA of Amazon, Apple, Facebook, and Google, 1st edn. Penguin Random House UK
Gillooly P (2016) Don't quit social media. Put it to work for your career instead. The New York Times. https://www.nytimes.com/2016/12/03/jobs/dont-quit-social-media-put-it-to-work-for-your-career-instead.html?searchResultPosition=4. Accessed 07 July 2023
Goel V, Perlroth N (2016) Yahoo says 1 billion user accounts were hacked. The New York Times. https://www.nytimes.com/2016/12/14/technology/yahoo-hack.html. Accessed 07 July 2023
Goodson S, MarketShare (2012) If you're not paying for it, you become the product. Forbes. https://www.forbes.com/sites/marketshare/2012/03/05/if-youre-not-paying-for-it-you-become-the-product/?sh=6b0537b85d6e. Accessed 07 July 2023
Gray ML, Suri S (2019) Ghost work: how to stop Silicon Valley from building a new global underclass, 1st edn. Harper Business
Hansell S (1999) In wired world, much is free at click of a mouse. The New York Times. https://www.nytimes.com/1999/10/14/business/in-wired-world-much-is-free-at-click-of-a-mouse.html. Accessed 07 July 2023
Hermalin BE, Katz ML (2018) What's so special about two-sided markets? In: Guzman M (ed) Toward a just society, Joseph Stiglitz and twenty-first century economics, 1st edn. Columbia University Press, pp 111–130
Huettinger M, Boyd JA (2020) Taxation of robots – what would have been the view of Smith and Marx on it? Int J Soc Econ 47(1):41–53
IBM (2022) Cost of a data breach report. https://www.ibm.com/security/data-breach. Accessed 07 July 2023
Janeček V (2018) Ownership of personal data in the Internet of Things. Comput Law Secur Rev 34: 1039–1052

Kerr I, Szilagyi K (2016) Asleep at the switch? How killer robots become a force multiplier of military necessity. In: Calo R, Froomkin MA, Kerr I (eds) Robot law, 1st edn. Edward Elgar, pp 333–365

Kesan JP, Hayes CM, Bashir MN (2016) A comprehensive empirical study of data privacy, trust, and consumer autonomy. Indiana Law J 91(2):267–352

Kim NS, Telman JDA (2015) Internet giants as quasi-governmental actors and the limits of contractual consent. Missouri Law Rev 80(3):723–770

Lanier J (2013) Who owns the future? 1st edn. Simon & Schuster

Laudon KC (1996) Markets and privacy. Commun ACM 39(9):92–104

Leack J (2018) Stop wasting time on social media and learn one of these skills instead. Mandatory. https://www.mandatory.com/living/1478185-social-media-wasting-time. Accessed 23 Aug 2022

Litman J (2000) Information privacy/information property. Stanf Law Rev 52(5):1283–1313

Madrigal AC (2012) Reading the privacy policies you encounter in a year would take 76 work days. The Atlantic. https://www.theatlantic.com/technology/archive/2012/03/reading-the-privacy-policies-you-encounter-in-a-year-would-take-76-work-days/253851/. Accessed 07 July 2023

Malcolm F, Roehrig P, Pring B (2017) What to do when machines do everything: how to get ahead in a world of AI, algorithms, bots, and big data, 1st edn. Wiley

Mankiw NG (2017) Principles of economics, 8th edn. Cengage Learning

McQuinn A (2018) No, internet users are not paying with their data. Inside Sources. https://insidesources.com/no-internet-users-not-paying-data/. Accessed 07 July 2023

Meisel W (2013) The software society: cultural and economic impact, 1st edn. Trafford Publishing

Menger C (2007) Principles of economics, reprinted edn. Ludwig von Mises Institute

Molina-Jimenez C, Al Nakib HD, Song L et al (2020) A case for a currencyless economy based on bartering with smart contracts, ArXiv. https://arxiv.org/abs/2010.07013. Accessed 07 July 2023

Morton P (2014) Profit splits and value chain analysis key to addressing transfer pricing issues in the digital economy. International Tax Review. https://www.internationaltaxreview.com/article/2a6a1rtfrvu75ml4p19mo/profit-splits-and-value-chain-analysis-key-to-addressing-transfer-pricing-issues-in-the-digital-economy. Accessed 07 July 2023

Moynihan Q, Asenjo A (2019) Facebook quietly ditched the 'it's free and always will be' slogan from its homepage. Business Insider. https://www.businessinsider.com/facebook-changes-free-and-always-will-be-slogan-on-homepage-2019-8?r=US&IR=T. Accessed 07 July 2023

OECD (2018) Tax challenges arising from digitalisation – interim report 2018: inclusive framework on BEPS, OECD/G20 Base Erosion and Profit Shifting Project. OECD Publishing

Omaar H (2021) No, data economy is not a barter economy. Center for Data Innovation. https://datainnovation.org/2021/09/no-the-data-economy-is-not-a-barter-economy/. Accessed 07 July 2023

Perlroth N (2013) Target struck in the cat-and-mouse game of credit theft. The New York Times. https://www.nytimes.com/2013/12/20/technology/target-stolen-shopper-data.html?_r=0. Accessed 07 July 2023

Posner EA, Weyl EG (2019) Radical markets: uprooting capitalism and democracy for a just society, 1st edn. Princeton University Press

Purtova N (2015) Illusion of personal data as no one's property. Law Innov Technol 7(1):83–111

PwC (2021) The potential impact of artificial intelligence on UK employment and the demand for skills, a report for the Department for Business, Energy and Industrial Strategy. https://assets.publishing.service.gov.uk/government/uploads/system/uploads/attachment_data/file/1023590/impact-of-ai-on-jobs.pdf. Accessed 07 July 2023

Rees C (2013) Tomorrow's privacy personal information as property. Int Data Privacy Law 3(4): 220–221

Rysman M (2009) The economics of two-sided markets. J Econ Perspect 23(3):125–143

Satariano A, Confessore N (2018) Cambridge Analytica's use of Facebook Data Broke British Law, watchdog finds. The New York Times. https://www.nytimes.com/2018/11/06/technology/cambridge-analytica-arron-banks.html?searchResultPosition=5. Accessed 07 July 2023

Shaw DW (1992) Searching for the opportunity cost of an individual's time. Land Econ 68(1): 107–115

Singer N (2012) Mapping and sharing, the consumer genome. The New York Times. https://www.nytimes.com/2012/06/17/technology/acxiom-the-quiet-giant-of-consumer-database-marketing.html. Accessed 07 July 2023

Singh PJ, Vipra J (2019) Economic rights over data: a framework for community data ownership. Development 62:53–57

Solove DJ (2013) Privacy self-management and the consent dilemma. Harv Law Rev 126:1880–1903

Swinnen KK (2020) Ownership of data: four recommendations for future research. J Law Prop Soc 5:139–175

The US Federal Trade Commission (2014) Data brokers, a call for transparency and accountability. https://www.ftc.gov/system/files/documents/reports/data-brokers-call-transparency-accountability-report-federal-trade-commission-may-2014/140527databrokerreport.pdf. Accessed 07 July 2023

Thimmesch A (2016) Transacting in data: tax, privacy and the new economy. Denver Law Rev 94(1):145–194

Vanderbilt T (2013) The science behind the Netflix algorithms that decide what you'll watch next. Wired. https://www.wired.com/2013/08/qq-netflix-algorithm/. Accessed 07 July 2023

Varian H (2018) Use and abuse of network effects. In: Guzman M (ed) Toward a just society, Joseph Stiglitz and twenty-first century economics, 1st edn. Columbia University Press, pp 227–239

Weeks DC (2015) Netflix: integrating spark at petabyte scale. O'Reilly. https://www.oreilly.com/library/view/strata-hadoop/9781491928004/video228081.html. Accessed 07 July 2023

Wlosik M (2022) What is a data broker and how does it Work? https://clearcpde.cc/blog/what-is-data-broker. Accessed 03 Aug 2022

World Economic Forum (2011) Personal data: the emergence of a new asset class. World Economic Forum

Chapter 4
Could Pay-for-Privacy Options and Data Labour Markets End the Unlimited Collection of Free Data?

Abstract This chapter analyses the potential implications of establishing a data labour market combined with pay-for-privacy options. Under this combined method, users will be able to protect their privacy by opting for a pay-for-privacy option or to obtain compensation directly from the data aggregator in exchange for sharing their data. Although founding such a market would constitute the best solution to solve the majority of the data market's current problems, it is not possible to implement it, at least for the moment being. This is due to several factors. To begin with, the different manners in which data are being collected and negative data externalities emerging from data collection would prevent such a system from being efficient. Also, to function properly the system has to be supervised by a competent authority, which may, on its own, give rise to further complications. Along with these problems, a data labour market necessitates a precise valuation of each piece of datum, which amounts to a nearly impossible task, at least for the moment being. In addition, a proper implementation of such a model would require a certain bargaining power of users, which is simply non-existent (for now). On top of all these difficulties, a micropayment system that remunerates users automatically cannot be implemented at this stage in time due to technical insufficiencies. Last but not least, a data labour market combined with pay-for-privacy options may produce the undesirable result of turning privacy into a luxury that needs to be afforded, disadvantaging, thereby, a certain part of the population in a constant manner.

4.1 Pay-for-Privacy Options Combined with a Data Labour Market: The Perfect Solution?

In its *step one*, the current functioning of the data market creates two major problems: even the most precautious users cannot protect their data in an efficient manner and users' data are constantly being collected without proper compensation by entities that are earning substantial profits on that very basis.

Among the several suggestions that have already been made to remedy these problems, the most prominent ones are *creation of pay-for-privacy options* and

X. Oberson, A. E. Yazıcıoğlu, *Taxation of Big Data*,
https://doi.org/10.1007/978-3-031-43372-6_4

creation of a data labour market. A pay-for-privacy option entails that users may choose to pay a certain amount of money instead of sharing their data while using a digital service. 'A data labour market', 'a personal data market', 'personal data economy' or simply 'a data market' all designate a market where users can monetise their data. The difference between a 'data labour market' and other 'data markets' resides in the ownership or control of data once they have been collected. If once collected data are deemed to be owned by data aggregators, then the relationship between users and data aggregators is analogous to that of an employee and employer, where the former has been actively engaged by the latter and whose time and inputs should be contracted and compensated, without, however, owning their 'work product', in the case at hand, their data.[1] For the purposes of this book, the term 'data labour market' is used as an umbrella term to simply designate a market where users monetise their data. By opting for this term, the authors are not intending to express an opinion about the ownership of data, a subject which remains beyond the scope of this contribution. The term 'data labour market' was chosen to avoid all terminological confusion, since the data that can be monetised by users are not limited to personal data and 'data market' is already used in this book to designate a larger concept.[2]

Creation of pay-for-privacy options and a data labour market are, in fact, interdependent and they should be implemented in a simultaneous manner to constitute a viable solution. As a matter of fact, availability of pay-for-privacy options necessarily entails that some users will not entertain such options and opt for accessing digital services for 'free'. The implemented solution will thus reveal to be only a partial remedy, as the digital barter will continue to exist. In the same vein, creation of a mandatory data labour market will have the effect of forcing all users to become labourers in the market concerned, without giving them the possibility to simply not share their data at all. Creating pay-for-privacy options along with a data labour market, however, will have the merit of allowing some consumers to pay for their privacy and some consumers to become data labourers. The digital barter will cease to exist and users attaching importance to their data will be able to protect them in an efficient manner.

In this 'combined' model:

- companies have the option of offering basic and additional upgraded services; however, they are under the obligation to provide a pay-for-privacy option, under which no data are collected from users (with the notable exception of data

[1]Aslam and Shah (2020), p. 22. The use of the term 'data labour market' does not entail that the relation between users and data aggregators should be legally qualified as an employee-employer relationship. It merely indicates that the two relations are analogous to a certain extent on a factual basis.

[2]The term 'data market' and similar terms are also used throughout this book to designate 'data labour market', in cases where that particular term was employed by the contributor whose idea is being explained in detail.

necessary to render the services concerned, but such data cannot be used for any other motives by companies); and

- users have the option between: (1) sharing their data and using basic digital services, with or without additional upgraded services, in exchange; and (2) opting for the pay-for-privacy option and using basic digital services, with or without additional upgraded services, without sharing their data (with the notable exception of data necessary to render the services concerned, but such data cannot be used for any other motives by companies).

Four different types of transactions can thus occur: (1) the company provides a basic and/or additional upgraded service(s), the consumer opts for the basic service and provides their data, the company remunerates the consumer for the difference between the value of the service and the value of the data (or *vice versa*); (2) the company provides basic and additional upgraded services, the consumer opts for one of the additional upgraded services in addition to the basic service and pays the price of that additional service along with providing their data, the company either pays the consumer for the difference between the value of the service and the value of the data or deducts the value of the data from the amount that should be paid by the consumer for the purchase of the additional service (or *vice versa*); (3) the company provides a basic and/or additional upgraded service(s), the consumer opts for the basic service and the pay-for-privacy option, the consumer pays the fee for the pay-for-privacy option, no data are collected from the consumer; and (4) the company provides basic and additional upgraded services, the consumer opts for one of the additional services on top of the basic service as well as for the pay-for-privacy option, the consumer pays the price for the additional service and the fee for the pay-for-privacy option, no data are collected from the consumer.

The combination of pay-for-privacy options and a data labour market would give users a choice between being remunerated for their work as 'data labourers' and preventing companies from collecting their data. Such an approach seems, *prima facie*, to constitute the perfect solution that would solve the two major problems occurring in *step one*. This initial observation is, unfortunately, not accurate for the moment being due to six different reasons.

First, implementing a pay-for-privacy option can impose a certain level of restriction on the unlimited collection of free data, but cannot put an end to that practice. This failure is based on two main factors: (1) different manners in which data are currently being collected by a variety of actors active in the market; and (2) negative data externalities. *Second*, to be effective, the correct implementation of the pay-for-privacy option should be controlled by the competent authorities, and it is not clear to what extent such control may be effectuated. *Third*, instating a pay-for-privacy option will necessarily amount to a disadvantageous treatment of a certain portion of the global population, who can afford to pay neither for the services nor for their privacy and who, thereby, will be forced to take part in the data labour market as labourers. *Fourth*, both for pay-for-privacy options and the data labour market to function effectively, the value of data should be established in a quasi-exact manner. Such a precise value determination, however, cannot be effectuated at this stage in time. *Fifth*, for the data labour market to function, each user has to be identified and remunerated automatically by means of micropayments through a system put in place to that end. Not only the establishment of such a system is not technically possible for the moment being, but also the system itself may have the effect of compromising users' data. *Last but not least*, an efficient data labour market would require a comparable bargaining power of each party to the transaction. Due to the current dynamics of the data market, users' bargaining power is inexistent.[3] Implementing a data labour market would, therefore, not reveal to be effective, unless the data market undergoes a dramatic transformation, which does not seem to be feasible in the near future.

The following sections examine these different aspects one by one to conclude that the best solution to remedy to the two fundamental problems materialising in *step one* cannot possibly be implemented at this point in time.

4.2 A Pay-for-Privacy Option Cannot Eradicate 'the Unlimited Collection of Free Data'

As adequate as it may seem to guarantee the privacy of user data, imposing a mandatory implementation of pay-for-privacy options on data aggregators is not able to eradicate the current practice of 'unlimited collection of free data'. Such lack of ability is caused by two features of the data market, namely (1) different manners in which data are currently being collected by a variety of actors active in the market; and (2) negative data externalities.

4.2.1 Pay-for-Privacy: Not an Option Implementable by All Data Aggregators

Pay-for-privacy system may reveal to be ineffective simply due to the different manners in which data are currently being collected. While using the services offered by a multi-sided platform or a similar company providing digital services, a consumer may efficiently benefit from a pay-for-privacy option offered by that platform or company. In that particular scenario, the consumer is aware of entering into a transaction and is able to consider their options with regard to the transaction concerned.

Nonetheless, as mentioned several times, data brokers and other data aggregators, such as Oracle and Acxiom, are collecting data by different means, and most of the time without the knowledge of users. The very data that users are aiming to protect by means of pay-for-privacy options, will therefore continue to be collected by other

[3] Fadler and Legner (2022), p. 124; Jurcys et al. (2020), p. 6.

companies, who cannot present users with any pay-for-privacy options due to the simple fact that they are not entering into a transaction (or any type of relations) with users when collecting their data. Whether such data collection should be deemed appropriate certainly constitutes a fundamental question, but it remains beyond the scope of this contribution. Enactment of a legislation, which would prevent such companies from collecting and aggregating users' data, may guarantee user privacy and render pay-for-privacy options a meaningful alternative.

In the meanwhile, imposing pay-for-privacy options on only some data aggregators can obviously not achieve the purpose of ensuring user privacy. Such a method is not reasonable from a policy point of view either. Imposing a limit on only some data aggregators would distort the data market and grant an undue advantage to other data aggregators; and, thereby, would risk undermining user privacy to an even greater extent.

4.2.2 Negative Data Externalities

By sharing their data, an individual compromises not only their own privacy but also the privacy of other individuals. This phenomenon is referred to as a *negative data externality*. In less scientific terms, it may also be referred to as 'data pollution'.

Even though they may not grasp the exact extent—which would require thorough scientific research—users seem to be aware of negative data externalities. As a matter of fact, individuals entering into data-intense environments online reported to suffer from a certain degree of anxiety over the impact of the potential exposure of their private information.[4] The caution aroused by this sentiment is, however, often defeated by the anticipation that, one way or another, their information would be exposed anyway by the actions of others.[5]

A negative data externality may occur due to the data shared by an individual's personal contacts. Users who are sharing information are compromising, on top of their own privacy, the privacy of other individuals whose information is correlated with theirs.[6] Research conducted by Bagrow et al. establishes that one can profile an individual from their available social ties even when that individual forgoes the platform completely. The researchers demonstrate that as few as eight to nine of an individual's contacts are sufficient to obtain predictability comparable to that of the individual alone, without requiring any data generated personally by the individual.[7]

For a negative data externality to occur, a personal contact between individuals is not necessary. Individuals sharing data about their own behaviour, habits and preferences, are, in fact, generating considerable information about the behaviour,

[4] Ben-Shahar (2019), p. 118.

[5] Ben-Shahar (2019), p. 118.

[6] Acemoglu et al. (2019), p. 1.

[7] It is important to note that the study was conducted on Twitter. See, Bagrow et al. (2019).

habits and preferences of not only their personal contacts (i.e., friends, family, colleagues etc.) but also of other people with similar characteristics (like a high-school graduate man in his mid-twenties living in Milano).[8] Accordingly, when a subset of users are willing to part with their data, which are informative about other users whose value of privacy is high, inefficiencies arise. For the purposes of pay-for-privacy options, such inefficiencies constitute a 'deal-breaker'. The data of consumers choosing pay-for-privacy options can be obtained, to a non-negligible extent, by analysing and processing other individuals' data.

As a remedy, Acemoglu et al. suggest '*de-correlation*', which means to transform the data of different users in such a manner as to remove the correlation that is at the root of the negative data externalities. According to the contributors, such a scheme can potentially be put in place by inserting a trusted third-party mediator into the transaction. Instead of sharing their data with the platform, users will share their data with the mediator, who can either share these data with the platform or transform them before revealing them to the platform.[9]

While de-correlation can prevent information leakage about those who prefer not to share their data, it is not clear whether such a system can effectively be implemented. Problems include establishing a trusted third-party mediator—who will not share the data obtained for any other purposes—, controlling the operations of such a mediator to prevent non-mediated information sharing between platforms and users, as well as to be able to guarantee the data safety during all these different stages. Acemoglu et al. themselves also underline that open questions as to the implementation of the system remain and such a scheme could be too complex to implement.[10]

Moreover, even if a system that would allow de-correlation can effectively be put in place, such a system does not seem to be able to constitute an effective solution for negative data externalities occurring in relation to individuals sharing similar characteristics. In that particular context, what is being compromised is not a specific piece of information about one specific person, which can be de-correlated by a trusted third-party, but the common traits of individuals sharing similar characteristics.

4.3 Necessity of Control by the Competent Authorities

Pay-for-privacy options can easily become an 'empty shell', if companies employ difficult-to-understand terms of service, which grant them the right to collect and/or store certain types of data, or if companies do not honour their responsibilities provided under their own terms of service. To ensure the efficient implementation

[8] Acemoglu et al. (2019), p. 1.

[9] Acemoglu et al. (2019), p. 34.

[10] Acemoglu et al. (2019), p. 35.

of a pay-for-privacy system, it would be necessary to control whether data aggregators are indeed abstaining from collecting any user data that are not directly relevant for the services they are providing as well as whether the data collected to render the services concerned are not being used for any other motives and are destroyed when no longer pertinent.

Such a control mechanism is far from being easy-to-implement. It would require detailed legislation and, most probably, the establishment of a competent authority in charge of effectuating the necessary controls and supervising the correct implementation of the legislation. A national legislation and a national competent authority cannot guarantee the relevant level of control and supervision, given the fact that data aggregators and users are often localised in different countries. It remains unclear to which extent such an audit can be conducted both from a legal and from a practical perspective.

4.4 Disadvantageous Treatment of a Certain Part of the Global Population

Albeit not being a difficulty inherent to the implementation of the system, a pay-for-privacy option, by its very nature, implies that some people will accept to 'sell' their privacy and some people will not. Residents of under-developed and developing countries as well as low-and-middle income earners are more likely to employ the option of commercialising their privacy. Fees for benefitting from pay-for-privacy options could form a significant economic burden for such consumers, who may be compelled to 'work as data labourers'. Pay-for-privacy options may thus have the effect of transforming privacy into a luxury that needs to be afforded. Moreover, the scarcity of data of the wealthier individuals may provoke a higher demand for such data and cause their value to increase in a substantial manner.

Establishing a pay-for-privacy option combined with a data labour market could, consequently, end up benefitting the wealthiest part of the global population, while placing other individuals in a position of disadvantage. Such an end-result is in direct conflict with the primary aim of that model: protecting users' interests.

4.5 Determining the Value of Data: A Quasi-Impossible Task

Establishing the exact value of data can easily form a textbook example of the adjective 'impossible'. In the current state of affairs, no one would argue to the contrary. For the moment being, determining the precise value of data faces serious limitations coming both from the functioning of the market and the available technical instruments. The current state of affairs may, however, be subject to change in a rather short period of time. Or not. While no one, at least not the authors of this book, can know when the currently reigning uncertainty will dissipate, two things are certain for the moment being: determining the exact value of data is a quasi-impossible task and without an accurate value of data, an efficient data market cannot be created.

4.5.1 *Creation of Data Labour Markets*

Data are currently treated as the capital of data aggregators. Some commentators argue that data should be considered as labour and that consumers providing their data should be seen as 'data workers', while some other commentators assert that a data labour market should be created.[11] A number of suggestions have already been made in view of creating a data labour market.

The earliest suggestion of establishment of a 'data market' seems to have been made by Laudon in 1996. Laudon suggests creation of a National Information Market, the functioning of which can be summarised in the following manner: (1) individuals would establish information accounts and deposit their information assets and information rights at a local information bank, which could be any local financial institution interested in taking part in the data market; (2) the local information bank would merge local depositor assets to form an information pool and transform that aggregated data into marketable information baskets; (3) the local bank would sell the baskets on a National Information Exchange; (4) buyers purchasing the baskets would obtain the right to use the information for commercial purposes for a limited period of time; (5) the amount obtained from the transaction would flow back to the local information bank and ultimately individuals' accounts would be credited (information-handling charges for the local banks, the Exchange as well as any brokers involved would be taken into consideration).[12]

In a similar vein, Lanier recommends the establishment of a 'humanistic information economy', in which users are remunerated by means of nanopayments for the content they create and each use of the data they generate.[13] This more recent proposition, which dates back to 2013, is based on the ability to trace the data and the content to the user who generated them. For example, in the context of an open-source computer code, the code would 'remember' the users who coded each line, and the users concerned would be sent nanopayments as part of code execution.[14] Within the humanistic information economy, users would also be remunerated when they contribute only passively through the use of their data; i.e., for results obtained

[11] See, for example, Arrieta-Ibarra et al. (2018), Posner and Weyl (2019), Laudon (1996) and Lanier (2013).

[12] Laudon (1996), pp. 99–101.

[13] Lanier (2013), pp. 250–256.

[14] Lanier (2013), p. 253.

by algorithms by means of the use of their data. To illustrate this point, Lanier gives the example of an individual meeting their future spouse on an online dating service. The algorithms that implement the dating service would take note of that marriage. Should that marriage reveal to be long-lasting, the algorithms would increasingly apply what seemed to be the correlations between the couple to matching other prospective couples. When some of these couples also get married, the correlations from the initial couple's case that were particularly relevant to the recommendations would be automatically calculated and the individuals forming the initial couple would get nanopayments as a result.[15] Lanier underlines that such calculations are already being made today, but the nanopayments are not effectuated.[16]

In a similar vein but a different format, a number of companies already put in place business models involving users' data, which can be regrouped into two categories, namely the data-insights model and the data-transfer model.[17] Companies following the data-insights model, such as Digi.me and Meeco, are providing users with platforms to control, compile, aggregate and obtain insights from their data, while also offering, in some cases, a marketplace for users to monetise their data through various means.[18] Companies using the data-transfer model, like Datacoup,[19] are providing a marketplace for users to transfer their data (or rights in the data) directly to the companies themselves or to unaffiliated third parties.[20]

Whether organised as a thorough ecosystem with a sophisticated payment method or as a business model followed by a number of companies, a data labour market that allows users to commercialise their data primarily relies on one factor: the ability to determine the value of data.

4.5.2 *Tackling the Impossible Task: Establishing an Approximate Value of Data*

User data constitute a heterogenous good and their value may depend on several factors. Enumerating all elements that can potentially produce an impact on the value of data not being feasible, only a number of features will be mentioned to serve as examples. The value of user data varies from one individual to another based on their economic profile, which encompasses their affluence (purchasing power),

[15] Lanier (2013), pp. 254–255.

[16] Lanier (2013), p. 255.

[17] Elvy (2017), p. 1392.

[18] Elvy (2017), p. 1392.

[19] Datacoup, a New York based start-up company, shut down its operations in November 2019. Currently, the company seems to conduct its activities as a part of ODE (which stands for Open Data Economy). For more information, see Datacoup (2022).

[20] Elvy (2017), p. 1392.

propensity to spend and consumption habits.[21] The value of user data may also depend on the intensity of the user's engagement with a service. More active users may generate significantly larger data volumes than others, creating, thereby, increasingly accurate user profiles that may make them more valuable to a company.[22] The value of the same user's data may change due to key life events, such as childbearing and a certain type of sickness.[23] The ultimate value of data can even be partially independent of their intrinsic value and depend on the data aggregator. The more users a data aggregator has, the more valuable the aggregate dataset collected for the purposes of inferring trends and preferences. The value of data tends to increase more than proportionately with the volume collected.[24]

Users can thus generate data of differing quantity and quality, which makes valuing data complex and highly context-dependent.[25] Raw data are even harder to value, given the fact that their economic usefulness depends on the individual capacity of their acquiring firm to distil relevant information from them and employ it efficiently.[26] It has been argued that while establishing an objective market value for data constitutes a challenging exercise, it forms a relatively easy task for data aggregators, which can proceed to an estimation based on *the revenues they can extract at a later stage* on the basis of that data.[27] Data aggregators can thus determine the value of the aggregated, and in the majority of cases even processed, data that they have, once they used them either for internal purposes or in the course of their commercial relations with third parties. This argument has to be considered with diligence. In some cases, data aggregators may not be in a position to be able to determine the value of the data on the basis of their final revenues from certain activities.[28] Data not being a physical commodity, tracing them from the point of extraction to a specific point of consumption may not be feasible.[29] In the course of their use, data are likely to be blended, processed and analysed for delivering a variety of goods and services in a range of different countries.[30] At times, profits may even be derived from data many years after they have been collected.[31] To sum up, determining the exact value of a datum belonging to a specific user at the moment of collection is extremely cumbersome, if not outright impossible.[32]

---

[21] Aslam and Shah (2020), p. 13.
[22] Aslam and Shah (2020), p. 14.
[23] Aslam and Shah (2020), p. 14.
[24] Aslam and Shah (2020), p. 14.
[25] Aslam and Shah (2020), p. 14.
[26] Aslam and Shah (2020), p. 14.
[27] Carrascal et al. (2011), p. 1.
[28] Aslam and Shah (2020), p. 15.
[29] Aslam and Shah (2020), p. 15.
[30] Aslam and Shah (2020), p. 15.
[31] Aslam and Shah (2020), p. 15.
[32] Lanier (2013), p. 252.

The value of data may also be subject to rapid fluctuations. Depending on the circumstances of the relevant market and their use, data can go through a long period of being almost useless, then become incredibly useful over a short or a long range and then lose their value once again. For instance, a company managing a digital platform providing subscription-based streaming services is likely to have substantial amounts of users' data at its disposal, allowing its algorithm to make movie suggestions. An additional datum may not have a consequential value for the purposes of that particular platform. However, should the same company decide to invest in the production of other services, such as creating avatars for the metaverse and other virtual reality services, the same piece of datum may suddenly gain considerable value, since it can be used to design and improve the 'new' services. Once enough data have been accumulated to render and improve the services concerned, data may, once again, lose their value.

A non-negligible hardship in the determination of data's value in the context of a data labour market is users' complete lack of knowledge on the value of their data. It does not seem reasonable to require companies to disclose all the potential uses of data, including data that cannot be characterised as personal data, to create a transparent environment for the determination of the approximate price of a datum that a user intends to sell on the market. Such a disclosure is also likely to be in conflict with the applicable national legislations, which safeguard the trade secret. In the same vein, it would not be reasonable to expect from users, who generally do not even know the exact manners in which their data are going to be employed, to determine an approximate value for the use of their data. Moreover, even if companies did disclose all the potential uses of data that they intend to make, it would still be extremely cumbersome for a user to determine the price tag of their datum. How much should a user charge for disclosing their preference for movies of Wes Anderson over movies of James Cameron? Is this piece of information more or less valuable, and by how much, than a general preference for drama over science fiction?

While neither users nor data aggregators can be considered as being 'at fault' for this information mismatch, a data labour market built on these premises could only be structurally imbalanced, and, thereby, inefficient. Users would systematically remain in a price taking position, without an opportunity to bargain or to make an informed choice on their data's price. Data aggregating companies, on the other hand, would systematically remain in the position of power in determining the prices and the conditions under which data sales are conducted. In the authors' opinion, relegating the task to determine the value of data to an independent third party, such as a designated governmental authority, does not constitute a plausible option either. Establishing even an approximate price would require, among others, an extremely detailed knowledge of the data labour market, data aggregators' internal processes—potentially along with some trade secrets—and the specificities of each user's data. Implementing such a system would not only cause several legal challenges, including the ones provoked by the predominant cross-border nature of such transactions, but also impose a substantial burden on governments.

Not surprisingly, pinpointing an approximate value of users' data on the basis of the available research and literature reveals to be quite a cumbersome task. The greatest majority of the relevant sources tend to merely indicate that data are hard to value. Nonetheless, a limited number of estimations of data's value do exist, albeit not being supported by scientific research.

According to an article of The Financial Times published in 2013, the financial worth of user data is extremely low.[33] In the article concerned, it is stated that general information about an individual, such as their age, gender and location is worth a mere USD 0.50 per 1000 individuals (thus USD 0.0005 per person). Since marketers are willing to pay more for certain traits, like being a potential house buyer, the price of certain information may be higher. For instance, information about auto buyers may be worth around USD 2.11 per 1000 persons; or accessing lists of people with specific health conditions or taking certain prescriptions may have a price tag of USD 0.26 per person. Overall, the article estimates the sum total of data for most individuals to less than 1 dollar; with the notable exception for users with a high level of social influence, whose data can be worth more than USD 3.[34] The numbers contained in the article of The Financial Times refer to prices of specific datums. An approximate value of the aggregate data belonging to a user cannot be estimated on the basis of the prices concerned. The price offered by Datacoup to purchase consumer data, when it launched its operations around the same time the article of The Financial Times was published (in 2012 to be precise), provides for a better approximation. The company offered to purchase consumer data for a fixed price of USD 8 per month.[35] A more recent, and a much higher, estimate can be found in an article of The New York Times published in 2017, according to which the approximate value of data amounts to USD 1000 per person per year and that number continues to rise.[36] On the basis of these evaluations, it can be observed that a user's data seem to be worth between USD 96 and USD 1000 per year (USD 8 and USD 83.3 per month), which gives an average of USD 548 per year. This amount relates to the period between 2012 and 2017 and it is not possible to verify its accuracy.

Past mergers and acquisitions may also shed some light on the approximate value of data.[37] For instance, when Microsoft acquired LinkedIn in December 2016, an approximate amount of USD 260 was paid by user. The other two examples that can be mentioned are the acquisitions of Instagram in April 2012 and of WhatsApp in February 2014 by Facebook (Meta). The per user amount was USD 20 for Instagram and USD 42 for WhatsApp.[38] Undoubtedly, these values are highly dependent on the circumstances under which the transaction has been carried out, such as the

---

[33] Steel (2013).

[34] Steel (2013).

[35] Elvy (2017), p. 1397.

[36] Madsbjerg (2017).

[37] Aslam and Shah (2020), p. 14.

[38] Aslam and Shah (2020), p. 14.

bargaining power of the parties and the timing of the transaction. Also, the values concerned do not exclusively refer to the value of user data, but to the overall value of the target company.[39] As a consequence, the value of data calculated on the basis of past mergers and acquisitions cannot constitute a viable benchmark for determining the value of data in an objective manner.

A study conducted by Carrascal et al., over a period of 2 months from mid-July to mid-September 2011, aimed to measure the value that users attach to their own data. The main findings of the study revealed that users valued, on average and per transaction, information related to their offline identity—consisting of age, gender, address and economic status—at EUR 25, information related to their browsing history at EUR 7, data referring to their interactions on social websites at EUR 12 and data referring to their interactions on finance websites at EUR 15.5.[40] The researchers also observed that the value attached by users to their data did not fluctuate on the basis of the number of pieces of information. They valued one piece of information and multiple pieces of the same information in an equal manner.[41] It can, therefore, be stated that users evaluated their data at an average of EUR 16 per transaction in 2011. It is important to underline that how users value their data inextricably depends on their preferences regarding data privacy, as well as the value they place on the service for which they are trading their data.[42]

4.6 Current Impossibility of the Implementation of a Micropayment System

Establishment of a data labour market will necessarily require keeping track of users as well as implementing a micropayment system. For example, within the National Information Market that he suggests, Laudon argues that users' data can be tracked by means of National Information Accounts (NIA). In that model, every participating citizen would be assigned an NIA with a unique identifier number and barcode symbol.[43] The NIA would allow users to trace their data. The amount generated by each relevant use would be credited to the user's NIA.[44] Similarly, to construct the humanistic information economy proposed by Lanier, every content created and piece of information shared by a user should be traced back to the user concerned. Additionally, a system of nanopayments should be put in place.

Non-negligible issues of implementation may occur in the context of a micropayment system. First, tracking every use of consumers' data will inevitably

[39] Goulder (2020), p. 560.

[40] Carrascal et al. (2011), p. 2.

[41] Carrascal et al. (2011), p. 2.

[42] Aslam and Shah (2020), p. 14.

[43] In the authors' opinion a QR code would be more efficient than a barcode symbol.

[44] Laudon (1996), pp. 100–101.

give rise to privacy issues. Currently, users' data are being collected, but they can be anonymised after a certain period of time. In a system that aims to track every use of data, a complete anonymisation is not likely to constitute a viable option. Even if technological improvements may render a system with complete anonymisation to work seamlessly in the future, a need to address the privacy issues would certainly continue to arise.

Second, creation of a micropayment system could reveal to be more challenging than it sounds from a technological standpoint. In fact, a number of start-ups tried to create systems of micropayments in the late 1990s and early 2000s. One of these efforts eventually became the well-known payment platform PayPal.[45] In its early years, the overhead costs of PayPal resulted in the system to be used only for large transactions. For a micropayment system destined to be employed in the context of social networking, the IoT and other similar services, where a substantial number of interactions are occurring in a swift manner, the costs could reveal to be even higher. Micropayments that would be effectuated to users for every single employment of their data could simply be too low to justify the costs of micropayment platforms.[46] Another problem inherent to the system could be its potential vulnerability to abuse, as demonstrated by Microsoft's experiment with paying users for their data. It was observed during the experiment that a large number of bots abused Microsoft's system and extracted substantial amounts of money.[47] Such a misuse would harm the entire ecosystem, since it will result in loss of funds, which will benefit neither the data aggregators, nor the users (or any other actors of the data market).

Smart contracts may form an efficient tool in lowering the costs and ensuring the security of a micropayment system.[48] With or without the use of smart contracts, however, a significant effort needs to be deployed to implement a well-functioning micropayment system and a rather long period of transition is necessary.

Last but not least, a micropayment system would not function if implemented by only one (or a limited number of) company(ies) and country(ies). For an efficient data labour market to exist, it is, in fact, necessary to use the same payment system, or compatible systems, on a global scale. This entails the implementation of a micropayment system by a large number of countries (the residents of which are subject to data collection) and data aggregators, which requires an international agreement, collaboration and effort. This necessity of reaching an international consensus adds yet an additional layer of difficulty to the implementation of a micropayment system and is likely to give rise, on its own, into a substantially long period of transition.

---

[45] Posner and Weyl (2019), pp. 71–72.

[46] Posner and Weyl (2019), pp. 71–72.

[47] Posner and Weyl (2019), p. 79.

[48] Pirkowski (2018).

4.7 Low Bargaining Power of Users

In a data labour market built on the foundation of the current system of data business, users can only become price takers, who would have to agree on the prices offered by data aggregators. The domination of the data market by a handful of data aggregators combined with users' inability to estimate the value of their data will inevitably result in a—to say the least—low bargaining power of users.

While it cannot be excluded altogether that the prices offered by data aggregators will match the 'real' price of data, it is highly probable—especially from the perspective of multi-sided platforms—that the prices presented to users will not reflect the authentic prices. As a matter of fact, the existence of a limited number of entities competing for collecting users' data implies that if one of the entities starts paying more for certain types of data, competition among the services will quickly force all of the entities to pay more for all the data they are receiving.[49] A data labour market established upon the current dynamics of the data business would, thus, be based on the constructive and yet controversial assumption that data aggregators would be condoning the substantial costs that they will incur while collecting data and would disclose the real value of each piece of datum.

Several suggestions have been made to remedy to the low bargaining power of users in a prospective data labour market. The key solution seems to be the constitution of an 'entity' that can demonstrate a certain amount of power while negotiating with data collectors. One of the recommendations made to this end is the establishment of a 'data labour union'. Individual users may not have much bargaining power but a union filtering the access to users' data could establish itself in a position of power and could, to the extent relevant and necessary, even call for a powerful strike.[50] If a data market system akin to the one suggested by Laudon is implemented, 'the local information banks' gathering users' data and commercialising them, may represent users in the data market, just as they are currently representing individuals in financial markets.[51]

It would be more appropriate to determine the best method to safeguard the bargaining power of users when implementing the actual data labour market, to take into account the specificities of the market concerned. A significant challenge awaits!

[49] Posner and Weyl (2019), p. 78.

[50] Arrieta-Ibarra et al. (2018), p. 41.

[51] Laudon (1996), p. 102.

4.8 Intermediary Conclusion: The Best Solution Is Not Yet Available

Despite the current shortcomings, including technical limitations, in the authors' opinion the establishment of a data labour market with an option to pay-for-privacy for users who do not wish to take part in the market concerned would constitute the best solution for both the lack of privacy and the lack of adequate compensation for data.

A well-functioning and efficient data labour market combined with a pay-for-privacy option would allow the following: (1) users who do not wish to grant access to their data will not be under the obligation to do so, but they will not be refused access to digital services for this lack of will; (2) users who would like to take part in the data labour market will have the option to do so; (3) actors active in the data market will not have a public image of 'entities exploiting the consumers'; (4) users who do take part in the data market will be compensated adequately; (5) actors active in the data market will not be disadvantaged, due to the fact that they will merely be paying for the raw material that they are currently obtaining for 'free'.

With proper legislation and governmental supervision, a data labour market could function as any other market and benefit to all its participants. The key element is the necessity of time, which might reveal to be quite substantial, for the implementation and the transition to occur. In the meanwhile, there is a need for an intermediary solution to be implemented to safeguard the interests of the most vulnerable actor[52] of the data market: users. This 'second best' solution suggested by the authors, namely the data collection tax, is developed in Chap. 6.

References

Acemoglu D, Makhdoumi A, Malekian A, Ozdaglar A (2019) Too much data: prices and inefficiencies in data markets. NBER working paper series no. 26296, National Bureau of Economic Research

Arrieta-Ibarra I, Goff L, Jiménez-Hernández D et al (2018) Should we treat data as labor? Moving beyond "free". AEA Papers Proc 108:38–42

Aslam A, Shah A (2020) Tec(h)tonic shifts: taxing the "digital economy". IMF staff discussion note, IMF

Bagrow JP, Liu X, Mitchell L (2019) Information flow reveals prediction limits in online social activity, ArXiv. https://arxiv.org/abs/1708.04575. Accessed 07 July 2023

Ben-Shahar O (2019) Data pollution. J Leg Anal 11:104–159

Carrascal JP, Riederer C, Erramilli V, Cherubini M, de Oliveira R (2011) Your browsing behavior for a Big Mac: economics of personal information online, ArXiv. https://arxiv.org/abs/1112.6098. Accessed 07 July 2023

[52] The other 'most' vulnerable position in the data market belongs to microworkers. Throughout the following chapters, users and microworkers are classified as 'the most vulnerable parties' with regard to different relations and transactions.

Datacoup (2022) Homepage. https://datacoup.com/. Accessed 07 July 2023
Elvy S (2017) Paying for privacy and the personal data economy. Columbia Law Rev 117(6): 1369–1454
Fadler M, Legner C (2022) Data ownership revisited: clarifying data accountabilities in times of big data and analytics. J Bus Anal 5(1):123–139
Goulder R (2020) Should data extraction be taxed as a natural resource? Tax Notes International, 27 July 2020, pp 559–561
Jurcys P, Donewald C, Globocnik J et al (2020) My data, my terms: a proposal for personal data use licenses. Harv J Law Technol 33:1–14
Lanier J (2013) Who owns the future? 1st edn. Simon & Schuster
Laudon KC (1996) Markets and privacy. Commun ACM 39(9):92–104
Madsbjerg S (2017) It's time to tax companies for using our personal data. The New York Times. https://www.nytimes.com/2017/11/14/business/dealbook/taxing-companies-for-using-our-personal-data.html?searchResultPosition=1. Accessed 07 July 2023
Pirkowski M (2018) Data slavery and decentralized emancipation, Facebook, Google and the future of data ownership. Medium. https://medium.com/swlh/data-slavery-and-decentralized-emancipation-ec9cc1265608. Accessed 07 July 2023
Posner EA, Weyl EG (2019) Radical markets: uprooting capitalism and democracy for a just society, 1st edn. Princeton University Press
Steel E (2013) Financial worth of data comes in at under a penny a piece. The Financial Times. https://www.ft.com/content/3cb056c6-d343-11e2-b3ff-00144feab7de. Accessed 07 July 2023

Chapter 5
Current International Approach to 'Data Taxation'

Abstract This chapter aims to briefly examine a selection of the suggestions that were already made in relation to data taxation. The suggestions concerned can be regrouped into two main categories: the ones targeting a portion of the overall company revenues and the ones targeting the use and commodification of data. Are described, from the former group, the OECD's Two-Pillar Approach, digital taxes and three new nexus, namely, a new nexus built on the concept of 'sustained user relationships', a new permanent establishment nexus based on 'digital investment' and a new permanent establishment nexus based on 'free labour'. Are examined, from the latter group, data pollution tax, royalty on user-generated revenue, data-flow tax and data dependence tax to fund a data dividend.

The ever-growing data market gives rise to fascinating tax challenges. In fact, this observation holds true not only for taxation of data, but also for taxation of profits of 'highly digitalised businesses' active in the digital economy. In the face of the non-negligible challenges, various suggestions have already been made by a number of contributors and considerable work has been conducted at both international and regional levels. Most of this work does not concentrate merely on data but focuses on the more general and comprehensive subject of taxation of digitalised businesses.

Analysing all the work conducted in the field of taxation of the digital economy, or taxation of highly digitalised businesses, falls beyond the scope of this contribution, which aims to explore suitable methods for taxation of data. In line with this purpose, this chapter does not aim to provide an exhaustive and detailed review of all suggestions that were made in relation to taxation of the digital economy. Independent of this purpose, it is also subject to discussion to what extent such an exercise would have revealed to be useful for the readers, given the fact that 'taxation of the digital economy' constitutes, by far, the most popular international tax law subject of the last decade and has already been scrutinised extensively by a myriad of commentators, countries as well as international organisations.

From the perspective of data, the suggestions that have already been made to tackle the tax issues can be regrouped into two main categories: the ones targeting the use and commodification of data and the ones targeting a portion of the overall company revenues. The second category predominantly consists of proposals that

X. Oberson, A. E. Yazıcıoğlu, *Taxation of Big Data*,
https://doi.org/10.1007/978-3-031-43372-6_5

are aiming for providing solutions to the prevalent problems caused by digitalised business models, which also include data. They are offering solutions primarily based on the 'traditional' income taxation, hence their characterisation as 'the campaign to save income taxation' by Marian.[1]

The authors of this book are of the opinion that taxation of data would be best addressed by analysing *each of the four steps* separately and implementing a different solution specific to each step.[2] It is observed that the common tendency in the international tax law scene consists of *merging step one and step three* and suggesting a solution predominantly focalised on targeting overall company revenues—i.e., using corporate income tax, or a similar tax, to remedy both to the lack of compensation of users in the phases of data collection and generation and to the inability to tax highly digitalised businesses in source states. The authors do not agree with this general tendency and consider that for the purposes of *step one* a solution targeting the use and commodification of data should be adopted, while for the purposes of *step three* a suggestion targeting a portion of company revenues would constitute the best fit, due to the fact that in this step the problem is effectively linked to the distributive rules of double tax treaties (DTT) that prevent the levy of a corporate income tax in source countries. The authors' opinions on tax issues relating to each step are examined in detail in Chaps. 6–9.

To serve as a basis for the analysis conducted in the following chapters, the main suggestions that have been made over the last decade are briefly examined in this chapter. The chapter begins with a description of the suggestions targeting a portion of the overall company revenues. It proceeds with an examination of suggestions targeting the use and commodification of data. Canvassing all the work conducted and every proposal made in the field in detail is not only impracticable but also not relevant for the purposes of the main issues examined in this book. Consequently, only a part of the work and some of the proposals, which are deemed to be the most relevant ones by the authors within the data taxation framework, are delved into.

5.1 Suggestions Targeting a Portion of the Overall Company Revenues

Prior to examining different suggestions made, it is important to briefly summarise the rationale behind proposals targeting a portion of the overall company revenues. As already mentioned, most of such suggestions were made in a context much larger than taxation of data, namely taxation of highly digitalised businesses. As a direct consequence, a great part of the discussions effectuated in relation to these proposals does not relate to data. Such discussions are not relevant for the purposes of this

[1] Marian (2022), pp. 551–560.

[2] The reference is made to the 'steps' that are identified in Sect. 3.5.

book, and thereby not reproduced in this chapter, which aims to focus on data. For these reasons, is only examined as the main rationale, value creation by users.

Then, are examined the two most commonly used and/or the most widely discussed instruments targeting a portion of the overall company revenues, namely the OECD's Two-Pillar Approach and digital taxes. These instruments have already been analysed extensively by a number of contributors and institutions, to the point where all explanations will necessarily amount to the repetition of the same information. With this in mind, developments made on these two instruments are kept as concise as possible.

Lastly are explored, three new nexus proposals concentrating more heavily on the specificities of the data market and user-generated data. Are examined, namely, a new nexus built on the concept of 'sustained user relationships' by Becker and Englisch, a new permanent establishment (PE) nexus based on 'digital investment' by Schön and a new PE nexus based on 'free labour' by Collin and Colin, as well as by Parsons.

5.1.1 Value Creation by Users

5.1.1.1 Rationale Behind the Current System of DTTs

The system of DTTs was built upon an international compromise conceived by the League of Nations in the 1920s. Referring to an '*international compromise*' serves one important purpose: underlining that the common understanding reached was difficult to justify exclusively on the basis of solid theoretical developments and it involved a certain degree of 'arbitrariness' to preserve, as much as possible, the international balance.[3]

In a nutshell, the international compromise opted for the allocation of taxing rights for different categories of income between residence and source countries. Economic allegiance, which builds on the benefit theory, serves as a guiding principle for this allocation. Under the benefit theory, the taxpayer and the government are seen as economic actors exchanging consideration with each other; i.e., an exchange of taxation in return for public goods and services.[4] On this basis, the principle of taxation of economic allegiance indicates that the source state, where investment is made, can legitimately claim taxation on the grounds of services provided to the taxpayer. Subsequently, where a person has business activities in a state other than their residence state, the allegiance to the former state is considered to be more substantial than to the latter state. The residence state should also get its 'fair share', but, according to von Schanz this 'fair share' is significantly lower than the appropriate share to be granted to the source state: three-quarters of the income

[3] Bruins et al. (1923).

[4] Elliffe (2021), p. 18.

earned by the business activities in question should be allocated to the state of source, and only the residual one quarter should be granted to the state of residence.

The concept of economic allegiance, along with administrative concerns of effective and efficient tax collection, paved the way for the current concept of PE, which sets a threshold of physical presence in the source country.[5] In its current version, PE acts as an administrative or compliance threshold,[6] which aims to ensure that taxes are paid where the enterprise would typically avail itself to a significant degree of public infrastructure and other public goods provided by the country, and where it would simultaneously be visible and accessible for tax authorities.[7]

Highly digitalised enterprises involved in the data market do not need to have a physical presence in countries with respect to which their services are rendered. This lack of a physical presence translates into the non-existence of a PE, and therefore into the absence of the right to submit such enterprises to tax in the source country. Hence, the necessity to find an appropriate solution in order to be able to tax a portion of the business profits in source countries.

5.1.1.2 Taking into Consideration 'the Value Created by Users'

One of the three proposals that were discussed during the earlier stages of the OECD's work on the digital economy, but were ultimately not retained, was the *user participation proposal*.[8] The user participation proposal was founded on the concept that soliciting the active and sustained engagement of users is a critical component of value creation[9] for certain highly digitalised businesses.[10] In this respect, the OECD underlined that the activities and participation of the users contribute to the creation of the brand, the generation of valuable data, and the development of a critical mass of users, which helps to establish market power, especially for the purposes of social media platforms, search engines and online marketplaces.[11] In other words, users are actually contributing to value creation by taking part in the production process, hence the new concept of 'prosumers'.[12]

[5] See Sect. 8.2.1 for a detailed analysis of the current PE thresholds.

[6] Cui (2019), (electronic version).

[7] Becker and Englisch (2019), p. 162.

[8] Along with the marketing intangibles proposal and the significant economic presence proposal. See OECD (2019), § 16–68.

[9] The concept of value creation is rather vague and its contours are flexible. On this point, see, for example, Elliffe (2021), pp. 34–41; Hey (2018), p. 203; and Das (2020), p. 135. It is often argued, with reason, that the flexibility of the concept of value creation facilitated reaching an international consensus to a great extent. See, for example, Becker and Englisch (2019), p. 162. This flexible and subjective nature of the term, however, bears the risk of bringing chaos and inconsistency into the international tax arena. See, for example, Das (2020), p. 142; and Herzfeld (2017), pp. 18–21.

[10] OECD (2019), § 18.

[11] OECD (2019), § 18–19.

[12] Becker and Englisch (2019), p. 166.

'The active and sustained engagement of users' mentioned by the OECD, should not be interpreted as denoting that users can only create value when they actively participate in the production process. The OECD regroups user participation into two broad categories: active and passive.[13] As indicated by the name of the categories in a quite straightforward manner, active participation is characterised by the existence of a direct activity that necessitates a certain amount of effort by the user (such as tagging, writing reviews and creating content); whereas passive participation does not involve a direct activity of the user, with the notable exception of some active steps that have to be taken in any case to make use of the digital services concerned, such as downloading an app, using a particular device and providing consent for user data to be collected.[14] The OECD illustrates the passive user participation by means of 'data collection', which is "[...] a by-product of the Internet and is gathered without the user's direct involvement or active transmission of data", typically by means of cookies capturing browsing activity.[15]

The crux of the tax challenge is, as stated by the OECD, that "This value generated by user participation is not captured in user jurisdictions under the existing international tax framework, which focuses on the physical activities of a business itself in determining where profits should be allocated and the extent of the taxing rights of user jurisdictions. This results in businesses being able to generate significant value from a jurisdiction with a significant and engaged user base (user jurisdiction) without the profits they derive from that value being subject to local tax".[16]

The OECD's assertion that both active and passive user participation contributes to the value creation of highly digitalised businesses has been subject to debate. Given the blurry contours of the term 'value creation', this controversy cannot be solved with ease. For instance, according to Becker and Englisch, it is not possible to argue that users are creating value in cases where their role is limited to the mere acquiescence of being observed to allow the collection of data.[17] The contributors contend that the raw data are technically produced by businesses in such cases, which must take action to collect that data. The contributors further advance their reasoning by indicating that "[...] just not preventing value creation from happening is not part of value creation itself".[18]

Should the approach followed by Becker and Englisch be adopted, the majority of the data collected from users would remain out of the scope of the value creation concept. This point is also underlined by the contributors themselves, who argue that the inadvertent creation of value for the enterprises merely by passively generating data and not providing anything for the business cannot be considered as being

[13] OECD (2018), § 147.

[14] OECD (2018), § 147–149.

[15] OECD (2018), § 149.

[16] See OECD (2019), § 20.

[17] Becker and Englisch (2019), p. 168.

[18] Becker and Englisch (2019), p. 168.

equivalent to actively contributing to the value creation process. In the same vein, such users cannot be assimilated into employees, who are providing their active effort for the business. Under this point of view, are only considered to be 'truly valuable' user contributions in the form of microwork, improvement of non-commercialised goods (like Wikipedia) and crowd participation.[19]

The line of reasoning advanced by Becker and Englisch demonstrates the very reason for which *step one* and *step three* should not be examined in conjunction for the purposes of data taxation. It is clear that data generated by users are valuable and justify compensation. Trying to incorporate this exogenous value into the companies active in the digital economy, however, gives rise to unnecessary complications and causes the debate to largely 'miss the point'.

5.1.2 The OECD's Two-Pillar Approach

Undisputedly, the most significant and comprehensive project aiming to address the tax challenges posed by the digitalisation of business has been, and continues to be, effectuated by the OECD. To be able to reach a multilateral solution, which would constitute a revision of the compromise found in the 1920s, the OECD has undertaken two substantial work streams, which were termed as 'pillars', presumably to denote that they are significant and fundamental to the international tax system.[20] The work conducted by the OECD, and its end result 'the Two-Pillar Approach', is not specific to taxation of data, but aims to generally address the tax challenges that have arisen following the 'heavy' digitalisation of business.

The main purpose of *Pillar One* is to adapt the international income tax system to new business models through modifications effectuated in the profit allocation and nexus rules applicable to business profits.[21] To this end, Pillar One expands the taxing rights of source states (which consist of, in the context of some business models, jurisdictions where the users are located) in cases where there is an active and sustained participation of a business in the economy of that jurisdiction.[22] To fulfill this purpose, Pillar One was designed to consist of new nexus rules, eliminating the requirement to have a physical presence in a jurisdiction in order to establish a taxing right and of new profit allocation rules, formed by a three-tier combination,[23] namely Amount A, Amount B and Tax Certainty.[24]

It can be said that *Amount A*, which is focused on providing a solution to the existing problem in the international tax regime that has been highlighted by

[19] Becker and Englisch (2019), p. 171.
[20] Elliffe (2021), p. 167.
[21] OECD (2020b), § 6.
[22] OECD (2020b), § 6.
[23] Elliffe (2021), p. 176.
[24] OECD (2020b), § 7.

digitalised businesses such as multi-sided platforms, forms the most controversial component of Pillar One.[25] Amount A aims to grant a new taxing right to source states[26] over a share of residual profit calculated at a multinational group (or segment) level.[27] This new taxing right will co-exist with the current nexus and profit allocation rules.[28] In-scope companies are designated as multinational enterprises (MNE) with global turnover above EUR 20 billion and profitability[29] above 10% calculated using an averaging mechanism.[30] Extractive industries and regulated financial services are excluded from Amount A's scope of application.[31] Market jurisdictions eligible to receive Amount A are to be determined by means of the new nexus rules, establishing a threshold of at least EUR 1 million in revenue derived from the jurisdiction concerned.[32] With an aim to balance the interests of smaller jurisdictions, in particular developing economies, the nexus threshold is set at EUR 250,000 for jurisdictions with a GDP of less than EUR 40 billion.[33] Eligible market jurisdictions are to receive 25% of the residual profits[34] of in-scope MNEs using a revenue-based allocation key.[35] Losses are to be carried forward.[36] Are also provided within the design of Amount A, a mechanism to eliminate double taxation, along with a marketing and distribution profits safe harbour.[37] A simplified administrative process is to be developed to minimise the complexity as well as costs and burden of filing and payment.[38] The re-allocation of profits provided by Amount A is expected to be applicable to around 100 companies, at least until the turnover threshold is lowered to enable a larger scope of application.[39]

Amount B consists of a fixed return for certain baseline marketing and distribution activities taking place physically in a source state, in line with the arm's length principle.[40] The implementation of Amount B intends to simplify the administration of TP rules for tax administrations and lower compliance costs for taxpayers as well as to enhance tax certainty and reduce controversy between tax administrations and

---

[25] Elliffe (2021), p. 177.

[26] Referred to as 'market jurisdictions' by the OECD.

[27] OECD (2020b), § 7.

[28] OECD (2020b), § 10.

[29] Profit before tax/revenue.

[30] OECD (2021), p. 1; OECD (2022b), p. 5.

[31] OECD (2021), p. 1.

[32] OECD (2021), p. 1.

[33] OECD (2021), p. 1; OECD (2020b), § 10.

[34] Residual profits are defined as profit in excess of 10% of revenue.

[35] OECD (2021), p. 2; OECD (2020b), § 10.

[36] OECD (2021), p. 2; OECD (2020b), § 10.

[37] OECD (2021), p. 3; OECD (2020b), § 10.

[38] OECD (2020b), § 10.

[39] OECD (2022a), Question 2, p. 1.

[40] OECD (2020b), § 7.

taxpayers.[41] For these reasons, Amount B tends to be acknowledged as a key deliverable of Pillar One.[42] In a nutshell, Amount B will standardise the remuneration of related party distributors that perform 'baseline marketing and distribution activities' in the source state.[43]

The last component, *Tax Certainty*, encompasses processes to improve tax certainty through effective dispute prevention and resolution mechanisms.[44] It consists of two building blocks: (1) dispute prevention and resolution for Amount A and (2) dispute prevention and resolution beyond Amount A.[45]

Pillar Two is designed to ensure that large internationally operating businesses pay a minimum level of tax regardless of the location of their headquarters or their jurisdiction of operation.[46] It achieves this purpose by means of a number of interlocking rules that seek to (1) ensure minimum taxation while avoiding double taxation or taxation where there is no economic profit; (2) cope with different tax system designs by jurisdictions as well as different operating models by businesses; (3) ensure transparency and level playing field; and (4) minimise administrative and compliance costs.[47] Pillar Two consists of two interlocking domestic rules (the so-called Global anti-Base Erosion—GloBE—rules) and a treaty-based rule.[48]

The GloBE rules consist of: (1) an Income Inclusion Rule, which imposes top-up tax on a parent entity in respect of the low taxed income of a constituent entity; and (2) an Undertaxed Payment Rule, which denies deductions or requires an equivalent adjustment to the extent the low taxed income of a constituent entity is not subject to tax under an Income Inclusion Rule. The GloBE rules are applicable to MNEs that meet the EUR 750 million threshold as determined under BEPS Action 13.[49] Countries may, however, choose to apply the Income Inclusion Rule to MNEs headquartered in their territory even if they do not meet the threshold. Are not subject to the GloBE rules, a number of entities, such as government entities, international organisations and non-profit organisations.[50] The minimum tax rate used for the purposes of GloBE rules is determined as 15%.[51]

The treaty-based rule of Pillar Two, the *Subject to Tax Rule*, allows source jurisdictions to impose limited source taxation on certain related party payments

---

[41] OECD (2020b), § 11.

[42] OECD (2020b), § 11.

[43] OECD (2020b), § 12.

[44] OECD (2020b), § 7.

[45] OECD (2020b), § 7.

[46] OECD (2020a), Box 1.1.

[47] OECD (2020a), Box 1.1.

[48] OECD (2021), p. 3.

[49] Country-by-Country Reporting, https://www.oecd.org/tax/beps/beps-actions/action13/. Accessed 07 July 2023.

[50] OECD (2021), p. 4.

[51] OECD (2021), p. 4.

subject to tax below the minimum rate of 9%.[52] This tax is creditable as a covered tax under the GloBE rules.[53]

The Two-Pillar Approach is expected to allow all types of economies (developing, emerging or with a higher GDP) to benefit from extra tax revenues.[54] Under Pillar One, taxing rights on more than USD 125 billion of profits are expected to be reallocated to source states on a yearly basis.[55] The global minimum tax of 15% is estimated to generate around USD 150 billion in additional global tax revenues annually under Pillar Two.[56]

The OECD's Two-Pillar Approach has been subject to some criticism, especially due to its complexity, limited scope of application and presumably modest amount of contribution to taxes that are collected by source countries.[57] A detailed analysis and discussion of the Two-Pillar Approach is beyond the scope of this book, concentrating on taxation of data. As it can be deducted from the brief summary of the reforms provided by the two Pillars, despite beginning the journey with the purpose of finding a solution to the tax challenges posed by the digital economy, the OECD finally took a much broader approach.

The same observation holds true from the precise perspective of data. Amount A, addressing—among others—the issues posed by the use of data, the contribution of users and the measurement of their value, goes significantly beyond the concept envisaged in the user participation proposal and 'the data/user challenge'.[58] The main question is, therefore, to ascertain whether the issues arising in data taxation should indeed be addressed by an instrument going fundamentally beyond the data/user challenge. In the authors' opinion, given its particularities, taxation of data would be best governed by tailor-made solutions prepared specifically to that end.

[52] OECD (2021), pp. 4–5.

[53] OECD (2021), p. 4.

[54] OECD (2022a), Question 3, p. 1.

[55] OECD (2022a), Question 3, p. 1.

[56] OECD (2022a), Question 3, p. 1.

[57] See, for example, Lammers (2020); Postler (2020), p. 86; Brauner (2022), p. 4; and Avi-Yonah et al. (2022), pp. 297–304. It must be noted that on 11 July 2023, 138 Members of the Inclusive Framework made an Outcome Statement on the Two-Pillar Approach. As per the Statement, Pillar Two is 'already a reality', with over 50 jurisdictions taking steps towards its implementation [OECD (2023), § 2]. However, for the moment being, the situation remains less clear for Pillar One, in respect of which the work is still on-going [OECD (2023), § 6–13].

[58] For a detailed explanation, see Elliffe (2021), pp. 223–226.

5.1.3 *Digital Taxes*

5.1.3.1 Overview

The remarkable expansion of the digital economy combined with the absence of the ability to levy a corporate income tax on the activities conducted by highly digitalised enterprises in source countries paved the way for the proliferation of 'digital taxes'. Not surprisingly, the core purpose of digital taxes is to allow for the exercise of taxing powers in relation to business activities conducted without a physical presence in the territory of the country.[59] 'Digital taxes' is an umbrella term used to designate several measures, with disparate modalities, adopted by different countries. In line with their most relevant features, the measures implementing a separate levy can be regrouped under three main categories.[60]

The *first category* comprises measures for which the tax base and rationale establish a rather close link with the profits of an enterprise.[61] The most relevant example of such taxes is the 'diverted profits tax' introduced, for example, in the UK in 2015.[62]

The *second group* of measures are the ones aiming for equalising the tax burden of enterprises that operate remotely with those that have a physical presence in the country (the so-called 'equalisation levies').[63] Such taxes are imposed only on payments made from the tax-imposing jurisdiction.[64] The first and the most commonly given equalisation levy example is the one implemented by India, one of the forerunner countries of the race to introduce a digital tax. Following the Right Florists decision, in which the Indian Income Tax Appellate Tribunal held that payments made to Google Ireland Limited and Overture Services Incorporated (i.e., Yahoo US) for online advertisements of a flower business could not be subject to withholding tax in the absence of a physical presence in India,[65] an equalisation levy was introduced in this country in 2016. The tax is levied at a rate of 6% on the gross amount a business located in India pays to a non-resident enterprise for the provision of online advertisement services. It is important to underline that the tax applies only when a business located in India makes a payment to a non-resident enterprise for the provision of online services. The location of users is not relevant. For instance, if a Sri Lankan manufacturer purchases advertisement space on Google

[59] Pistone and Ullmann (2021), p. 168.

[60] Amendments made to the income tax system, without adopting a separate levy, may also be referred to as a digital tax. On this point, see Pistone and Ullmann (2021), pp. 173–174.

[61] Pistone and Ullmann (2021), p. 169.

[62] Pistone and Ullmann (2021), p. 169.

[63] Pistone and Ullmann (2021), p. 170.

[64] Cui (2019), (electronic version).

[65] *ITO v. Right Florists Pvt. Ltd.*, Income Tax Appellate Tribunal Kolkata, 12 April 2013, 154 TTJ (KOL) 142.

to target consumers in India, the levy would not apply, regardless of the fact that the users are located in India.[66]

The *third group* of taxes are digital services taxes (DST), by means of which the gross revenue is submitted to tax in order to catch the value created on the territory of a country.[67] In contrast to equalisation levies, DSTs grant the taxing rights on the basis of the location of users, regardless of whether the users make any payments.[68] Constituting, by far, the most popular category of digital taxes, DSTs are examined in greater details in the following section.

5.1.3.2 Digital Services Taxes

Many countries, among which an important number of EU Member countries, introduced a DST.[69] Most notably, on 21 March 2018, the EU Commission proposed to introduce a DST at an EU level,[70] which would have consisted of an interim measure until the more comprehensive solution, namely the new PE nexus based on significant presence,[71] could be adopted.[72] It is important to highlight that DSTs have traditionally been perceived as interim measures bound to disappear as soon as an international solution has been implemented.

While national DSTs may diverge from each other to a certain extent, they have a number of common characteristics. They are targeting 'tech giant MNEs' that are exceeding a certain global[73] and domestic turnover threshold and that are considered to be under-taxed in the countries where users are located. DSTs are typically levied on gross revenues from digital services, linked to the participation of local users, at a fixed rate varying between 2% and 7% (most commonly 3%). By locating the taxable income at the place where the users of the taxable service are situated, DSTs are levied both on resident and on non-resident enterprises, as opposed to equalisation levies. For instance, under a DST regime, the tax is applicable both to a Canadian and a French manufacturer purchasing advertisement space on Google to target consumers in France. DSTs are payable by the taxable person supplying the taxable services. In general, it is incumbent upon the taxable person to notify the country levying the tax, most commonly by registering as a taxpayer, and to file a tax return.

[66] Cui (2019), (electronic version).

[67] Pistone and Ullmann (2021), p. 171.

[68] Cui (2019), (electronic version).

[69] See Alvarado (2021), table 2 for a summary.

[70] European Commission (2018b).

[71] European Commission (2018a).

[72] It is to be noted that the scope of the EU's DST proposal was narrowed down. The text was re-drafted and re-focused on revenues resulting only from the provision of digital advertising services, i.e., digital advertising tax. For further information on this point, see, ECOFIN (2019).

[73] The global threshold is typically EUR 750 million.

The traditional allocation of taxing rights between source countries and residence countries thus disappears in the context of DSTs. Instead of aiming to allocate the taxing rights to source countries, the idea under the DST regime is to allow the country where the economic rent[74] is located to tax that rent.[75]

Several *legal controversies* exist with regard to DSTs. One of the most important issues is to determine whether they fall under the material scope of application of DTTs as per article 2 of the OECD and UN Model Conventions. For some contributors, DSTs constitute *sui generis* hybrid taxes,[76] while for others they form 'disguised' income taxes.[77] Should DSTs be qualified as taxes on income, they would be covered by tax treaties and subject to the PE threshold, which would prevent, once again, taxation in the source country. In contrast, should DSTs be qualified as taxes other than taxes on income (or on capital), they would not be covered by the material scope of DTTs and, thereby, their application will not be subject to treaty limitations. The international consensus seems to tend towards considering that DSTs do not fall within the scope of application of DTTs.[78] This point of view may predominantly be motivated by safeguarding the international balance through the preservation of the measures implemented by certain countries. On this point, the authors rather agree with Jiménez, who states that such taxes "[. . .] are income taxes in disguise that pretend to have a life outside income tax treaties, with the clear goal of recognizing more taxing rights to market states and [. . .] they can cause treaty overrides".[79]

Another legal ground on the basis of which DSTs are criticised is their restricted scope, which has been interpreted by some commentators and authorities as giving rise to discrimination. Avi-Yonah et al. conducted a to the point analysis on this issue and expressed pertinent conclusions. The contributors underline that implementation of a DST attempts to solve problems arising through some companies' ability to avoid tax nexus in foreign countries based on their absence of physical presence and that companies already having a tax nexus in the countries concerned should, naturally, not be covered by such taxes.[80] In the light of this very fact, a claim of discrimination remains unconvincing, since it would amount to ignoring the origins of the phenomenon.[81] The extremely limited number of enterprises falling within the personal scope of application of DSTs, mostly composed of US 'tech giants', should not be construed as amounting to discrimination either. The reason behind US tech

---

[74] For the concept of economic rent, see Cui and Hashimzade (2019), pp. 29–30.

[75] Cui (2019), (electronic version).

[76] See, for example, Pistone and Ullmann (2021), p. 180. On this point, see also Kofler and Sinnig (2019), p. 135, footnote 237.

[77] See, for example, Jiménez (2018), p. 630.

[78] On this point, the OECD indicates that a case-by-case basis analysis would be more appropriate. See, OECD (2018), § 415–424.

[79] Jiménez (2018), p. 630.

[80] Avi-Yonah et al. (2022), pp. 320–326.

[81] Avi-Yonah et al. (2022), p. 323.

giants being predominantly affected by DSTs, is simply the lack of any real foreign competition in that particular area.[82] It is also worth mentioning that in its two long-awaited decisions, rendered on 3 March 2020, the CJEU accepted the validity of a Hungarian tax on the turnover of advertising services, including digital services, under the principle of freedom of establishment of the EU law.[83]

Potential conflict of DSTs with the World Trade Organisation (WTO) standards forms yet another debated legal issue.[84] The most fascinating legal issue, however, may be yet to materialise. While the conditions for the removal of existing DSTs and relevant similar measures along with the commitment of not to enact new DSTs and relevant similar measures are amongst the substantive features of the text of the Multilateral Convention on Amount A of Pillar One,[85] the approach of the countries and their actual action remain to be seen. The possibility of some countries showing reluctance to revoke their DST despite the implementation of an international solution cannot be excluded altogether.

On top of the legal controversies, *a number of economic concerns* are expressed in relation to DSTs. Most notably, it is often invoked that DSTs may result in excessive taxation of digital businesses and that, in practice, they will simply be passed onto the consumers.[86] Such concerns may have held water in the 'traditional economy', but they do not seem to be justified in the context of the digital economy.

In fact, a common feature of the highly digitalised businesses is to operate with very low, often negligible, marginal costs.[87] The process of digital service provision tends to be largely automated with little additional labour and other input from the platform company, once the platform is 'up and running'.[88] This entails that the revenue of a platform company from each additional transaction is almost identical to its marginal profit from the transaction.[89] A tax on revenue (DST) is roughly equivalent to a tax on marginal profits.[90] Consequently, in cases where zero marginal costs constitute a reasonable approximation, which would often be the case, the platform company's marginal production and pricing decisions should be the same under a revenue tax (DST) or a profit tax (corporate income tax).[91] As a result, DSTs are not likely to give rise to *any excessive taxation* of digital businesses.

---

[82] Avi-Yonah et al. (2022), p. 282.

[83] Cases C-323/18 *Tesco-Global Aruhazak Zrt.* and C-75/18, *Vodafone Magyarorszag Mobil Tavközlési Zrt.* of 03 March 2020.

[84] Kofler and Sinnig (2019), p. 137.

[85] OECD (2023), § 7.

[86] For further information see, Bauer (2018), pp. 5–7.

[87] Cui (2019), (electronic version).

[88] Cui (2019), (electronic version).

[89] Cui (2019), (electronic version).

[90] Cui (2019), (electronic version).

[91] Cui (2019), (electronic version).

In the eventuality where platforms have positive marginal costs, a part of the costs incurred due to the levy of a DST may effectively be passed onto the consumers.[92] For instance, it has been observed that the enforcement of sales and hotel taxes on transactions on AirBnB resulted in higher listing prices, with most of the burden of the taxes passed onto the consumers.[93] Such pass-through does, however, not necessarily constitute 'a bug' of the system, but rather an element of design.[94] As a matter of fact, whether, and in the affirmative to whom, the burden of the tax will be passed onto depends on the business model followed by a platform. In the case of AirBnB, the purpose is to subsidise the landlords and, therefore, passing the cost of the tax onto the guests is consistent with the functioning of the platform.[95] The situation would be different in the context of a platform subsidising users, such as Google and Facebook. The cost of a DST would either be absorbed by the platforms themselves or passed onto the advertisers, and not the users.[96] In a business model where users are the key component on which the whole system is built, it would not be a reasonable decision to pass any costs onto users and cause, thereby, a decrease in the number of users or the time spent on the platform.[97] Accordingly, it is possible to conclude that the *cost of a DST will not automatically be passed onto consumers*.

As argued by Cui, in a quite convincing manner, DSTs constitute efficient taxes that allow location-specific rents earned by digital platforms to be submitted to tax in the countries where they arise.[98] Nonetheless, they do cause a number of legal issues, especially from an international tax law point of view. In the authors' opinion, the efficiency of such taxes cannot counterbalance the legal controversies they give rise to. Such controversies are, indeed, as serious as to be considered "[. . .] a direct attack to the current international tax order as we know it".[99]

5.1.4 New Nexus Proposals

The new nexus proposals detailed below have the merit of concentrating specifically on the particular features of the data market and user-generated data. The nexus built on the concept of 'sustained user relationships' by Becker and Englisch and the new PE nexus based on 'digital investment' by Schön are mainly concentrating on investments made and activities undertaken by highly digitalised enterprises to create and sustain efficient local user networks in source countries. The new PE nexus based on 'free labour' by Collin and Colin, as well as by Parsons, finds its

[92] Cui (2019), (electronic version).

[93] Cui (2019), (electronic version).

[94] Cui (2019), (electronic version).

[95] Cui (2019), (electronic version).

[96] Cui (2019), (electronic version).

[97] Cui (2019), (electronic version).

[98] Cui (2019); see also Cui and Hashimzade (2019).

[99] Jiménez (2018), p. 630.

roots in users acting as 'digital labourers' to generate data for the highly digitalised enterprises, without receiving an adequate compensation.

Albeit having a number of differences, both from a theoretical and a practical perspective, it can be stated that all the new nexus suggestions are actually based on the existence of a 'dedicated' user base generating data in the source country. In our view, data generated by users should be compensated by means of a data collection tax (*step one*); a portion of profits derived from digital services, provided either to users in exchange of payment or to customers on the other side(s) of platforms, should be submitted to tax in source countries (*step three*); and finally sale or transfer of data and granting the right to use a database should be regulated under the aegis of the current provisions of DTTs and international tax law practice (*step four*). With this caveat in mind, the authors of this book are also arguing for the creation of a new PE nexus for the purposes of *step three*, as developed in Chap. 8, albeit without an emphasis as pronounced as the one contained in the following nexus proposals on the local user base.[100]

5.1.4.1 A New Nexus Built on the Concept of 'Sustained User Relationships'

A nexus built on the concept of 'sustained user relationships' (SURE) is put forward by Becker and Englisch. It is to be noted that the contributors revised their suggestions on a number of occasions. The explanations made in this section aims to provide a general overview of the proposed new nexus and may not reflect all the modifications that were made by the contributors in different versions of the proposal. Reference is made to 'early' and 'revised'[101] suggestions, where necessary.

Becker and Englisch argue that while users assuming the role of 'unconscious contributors' cannot be considered as creating value,[102] the sustained access to a local network of users may constitute a crucial driver for value creation by highly digitalised enterprises following certain business models.[103] In this line of reasoning, a SURE built via provision of digital services constitutes an 'often unique and valuable' source of value creation in the production process.[104] As a matter of fact, the concept of SURE may serve as a proxy for the potential to legally collect certain types of data from users and to exploit them commercially.[105] On this basis, a

[100]It is to be noted that the nexus built on the concept of 'sustained user relationships' was not expressly designated as a PE nexus by Becker and Englisch. Should such a nexus be effectively implemented, however, it would be reasonable to incorporate it in DTTs as a new PE nexus.

[101]It is to be noted that the revised suggestion used in this book has a third author, namely Schanz.

[102]Becker and Englisch (2019), p. 168. This aspect is further detailed above in Sect. 5.1.1.

[103]Becker and Englisch (2019), p. 167.

[104]Becker and Englisch (2019), p. 169.

[105]Becker and Englisch (2019), p. 169.

local user network, actively created by an enterprise in a country and exploited as a core resource for the generation of business profits, may establish a nexus and legitimise the allocation of profits generated through its exploitation to the country concerned.[106]

In their early suggestion, the contributors put forward three cumulative conditions for the formation of the local user network nexus in a source country. First, a stable user relationship should exist. Stability is determined by the recurring use of an electronic interface operated by the enterprise or of data-transmitting devices controlled by the enterprise concerned, by a permanent user identity or otherwise. Second, the local network, or other form of stable user base, needs to be employed by the firm for business purposes. The activities conducted by means of the network must notably exceed the mere existence of a customer base for exported goods or services. In fact, the creation and exploitation of a sustained user network should form part of the value creation process within the firm, rather than merely reflect the demand side of the market. Third, to ensure administrative efficiency, the user network should be of sufficient size (in terms of number of users or devices) and should reach a certain intensity of use.[107] In their revised proposal, the contributors rely on two main criteria to establish a taxable local user network nexus: (1) the business activity associated with the local source of data or user network must be a *repeated one* or a *permanent one*; and (2) the activity must be *qualitatively and quantitatively important*.[108] The contributors then make further observations and enumerate different criteria on the basis of the types of activities conducted, such as advertising and marketing, to determine the existence of the nexus.[109]

With regard to attribution of profits to the new local user network nexus, in their early proposal, Becker and Englisch suggest that several methods could be designated to address the issues arising in relation to different business models, instead of one-size-fits-all allocation rules.[110] For instance, regarding online targeted advertisement firms, the international partition of the tax may be effectuated on the basis of the number of advertisements displayed on the devices of the users who form part of the network, in order to assess the significance of data access for the purposes of a 'simplified and standardised' profit split.[111] In their revised proposal, the contributors opt for the application of the profit split method with a four-step residual analysis.[112]

[106] Becker and Englisch (2019), p. 167.
[107] Becker and Englisch (2019), p. 171.
[108] Becker et al. (2019), p. 12.
[109] See Becker et al. (2019), pp. 13–15.
[110] Becker and Englisch (2019), p. 171.
[111] Becker and Englisch (2019), p. 171.
[112] See Becker et al. (2019), pp. 18–32.

5.1.4.2 A New PE Nexus Based on 'Digital Investment'

Schön suggests a new 'digital investment PE' nexus, which is rather akin to the local user network nexus of Becker and Englisch. The contributor begins his analysis by underlining that the existence of a local customer base should be distinguished from the existence of a local digital presence. The former cannot give rise to a taxable nexus in the source country *per se*, while the latter can.[113]

The formation of a local digital presence is dependent upon a tangible or intangible investment in the source country. According to Schön, the mere fact that an enterprise is highly digitalised does not imply that the enterprise concerned cannot dedicate intangible investment to respective country-specific markets with an aim of achieving a better customer access through offering tailor-made solutions, developing brands, operating services centres, etc.[114] With regard to a local user base, Schön states that "[i]f it can be shown that an Internet firm has invested capital in a specific market in order to access a specific customer base, it becomes evident that this investment can give rise to taxing rights in the respective market country, not simply because there is a market with customers ordering goods or services, but because the company has invested into that market and expects a return on this investment".[115] The digital services rendered to consumers in the context of the digital barter are to be taken into consideration when assessing the 'digital investment', given the fact that such investment creates market access for profit-generating services provided to the customers on the other side(s) of the multi-sided platforms.[116]

Schön concretises his suggested nexus by establishing two thresholds needing to be passed: (1) country-specific expenditure for digital investment by the taxpayer exceeding a certain multi-year threshold;[117] and (2) the size of the local user base exceeding a certain threshold (set in terms of number of users, number of contracts or amount of revenue).[118]

For the purposes of profit attribution, Schön argues that it is possible to apply the methodology developed by the OECD in 'traditional' business areas. To do so, the country-specific investment, including tangible and intangible assets created to that effect, along with the functions performed within the firm as regards to this investment should be identified. Value of users may also form a part of profit allocation, not because they provide a customer basis, but for the fact that they indicate the value of the investment a firm has made.[119] The physical presence of 'significant people' in

---

[113] Schön (2018), pp. 288–289.

[114] Schön (2018), pp. 288–289.

[115] Schön (2018), p. 289.

[116] Schön (2018), p. 289.

[117] The threshold is set on a multi-year basis since specific investment in hardware, software and other assets tends to have a useful life of more than 1 year. Schön (2019), p. 28.

[118] Schön (2019), p. 28.

[119] Schön (2018), p. 290.

the relevant country should, however, not be taken into consideration during the profit allocation process, since the concept of digital investment is not related to physical presence.[120]

5.1.4.3 A New PE Nexus Based on 'Free Labour'

In the study conducted by the Task Force on Taxation of the Digital Economy in France in 2013, Collin and Colin contend that the notion of PE needs to be adapted to effectively capture the 'free labour phenomenon', which they describe as "[...] the economic activities generated in a given country by the voluntary collaboration of application users in the production process of a foreign company".[121] According to the contributors, an enterprise that provides a service in a country by using data collected through regular and systematic monitoring of users in that country could be deemed to have a virtual PE there.[122]

Regarding the profit attribution to the new virtual PE nexus, Collin and Colin underline that the capacity to make users 'work' and the capacity to collect data should be considered as the equivalent of an asset, which should be evaluated at a fair value and attributed to the virtual PE created in the source country.[123] This would allow to allocate the profits of the data aggregating enterprise between different countries on the basis of the locations of the conventional intangible assets, the business activity and the users, who contribute to profit generation through their activities combined with the data obtained from regular and systematic monitoring of the activities concerned. In this line of reasoning, the primary component to include in an efficient profit attribution system is the ability to obtain a declaration of profits reflecting the actual respective contributions of the factors of production located in different countries, such as the intangible assets (particularly software) that the PE uses, and pays for, in order to carry its own operations.[124]

Collin and Colin take note of one significant obstacle to the accurate determination of profits attributable to the new PE nexus: inflated royalty payments made to foreign companies. Should the PEs of highly digitalised businesses located in one country pay more than justified amounts of royalties to their group companies or other PEs situated in other countries for the use of intangible assets (such as intellectual property and algorithms), the profits attributable to the former country could be reduced in a significant manner.[125] As a remedy, the contributors suggest challenging deductions for royalties paid to foreign entities for the right to use an

[120] Schön (2019), p. 29.

[121] Collin and Colin (2013), p. 114.

[122] Collin and Colin (2013), p. 115.

[123] Collin and Colin (2013), p. 116.

[124] Collin and Colin (2013), p. 117.

[125] Collin and Colin (2013), p. 115.

intangible asset in cases where such payments were made in the pursuit of treaty shopping.[126]

In a similar—but less elaborate and mainly concentrated in the US domestic law—manner, Parsons recommends the creation of a new 'workforce PE nexus'. In the light of a thorough analysis primarily focusing on the US domestic law, the contributor concludes that if the users are qualified as 'digital laborers' and treated in the same manner as a 'traditional' workforce, their home countries, which are generally the source countries, may be able to levy a tax on the income attributable to their work under their domestic laws.[127] This source taxation would, however, be limited under the current DTT system, due to the contours of the concept of PE. To remedy to that issue, Parsons advocates the creation of a PE on the basis of the presence of a workforce alone.[128]

5.2 Suggestions Targeting the Use and Commodification of Data

Four suggestions targeting the use and commodification of data are examined in this section. Namely, data pollution tax by Ben-Shahar, royalty on user-generated value by Aslam and Shah, data-flow tax by Marian and data dependence tax to fund a data dividend by Feygin et al.

5.2.1 Data Pollution Tax

> The digital economy creates digital smog; the question is what to do about it.[129]

Ben-Shahar argues that "data pollution is to our century what industrial pollution was to the last one".[130] The contributor identifies three manners to tackle data pollution; namely establishing a predetermined set of production restrictions—the most common form of which is quantity caps and quotas—, pricing the social cost—i.e., levying a Pigouvian tax—, and designing a liability regime for data harms with a particular purpose of addressing inadvertent data emission problems.[131]

As put forward by Pigou, the market price of goods and services remains lower than their real costs, when manufacturers externalise some of the costs that are linked

[126] Collin and Colin (2013), p. 120.
[127] Parsons (2022), p. 1831.
[128] Parsons (2022), p. 1831.
[129] Ben-Shahar (2019), p. 118.
[130] Ben-Shahar (2019), p. 106.
[131] For a summary of the three suggestions, see, Ben-Shahar (2019), pp. 108–110.

to production instead of bearing the full burden.[132] The externalised costs are borne by others. The rules and regulations in force, or more accurately their absence, may allow, to a certain extent, such negative externalities, like environmental and health damages.[133] Levying a tax, commonly referred to as a '*Pigouvian tax*' (or, under a more popular terminology, '*sin tax*'), may ensure the internalisation of the external costs to a certain degree.[134] The Pigouvian tax suggested by Ben-Shahar intends, thus, to internalise the costs linked to 'data pollution'.[135]

As also underlined by the contributor, it is quite possible that the benefits from data-driven personalisation far exceed the negative impact of data-driven discrimination and that it would be more appropriate to analyse 'data greens' rather than data pollution. That being said, the benefits of data are often appropriated and internalised: firms creating such benefits have the incentives and technical tools to commercialise and monetise them (before competition dissipates such gains). The negative externalities, in contrast, remain orphan. They affect groups too broad and dispersed and cause injuries that are too abstract for private remedies to be effective.[136]

Data pollution tax would be levied when the transaction takes place, which corresponds to the time of data collection. For instance, in the context of a purchase transaction effectuated between a consumer and a retailer by means of an online platform, the tax would be paid when the consumer is effectuating the payment.[137]

Establishing the rate of the data pollution tax constitutes a significant challenge. The noble purpose of internalising the externality meets a non-negligible obstacle: determining the size of externality. In an ideal world, the tax charged on polluters would be equal to the external costs.[138] However, measuring external costs with accuracy is difficult enough even when dealing with tangible externalities, such as damages to individuals and structures.[139] The negative externalities produced by data are even harder to compute. As underlined by Ben-Shahar, "until harms occur, it might be impossible to predict which data practices would be harmful, let alone their severity".[140] Recognising this difficulty, the contributor does not develop issues relating to the determination of the tax rate in detail. He argues that to the extent where administrative costs are not too large, some form of data tax, even if it is a

[132] See, Pigou (1932).

[133] For an overview of negative data externalities, see Sect. 4.2.2.

[134] For more detailed information, see Pigou (1932).

[135] Ben-Shahar (2019), p. 140.

[136] Ben-Shahar (2019), p. 115.

[137] Ben-Shahar (2019), p. 139. Although Ben-Shahar does not expressly state that the tax would be due at the moment of payment, that is the only timing that seems plausible in the light of the explanations given by the contributor.

[138] Laudon (1996), p. 98.

[139] Laudon (1996), p. 98.

[140] Ben-Shahar (2019), p. 139.

nominal one, is justified.[141] Nevertheless, the contributor makes certain suggestions that should ideally be taken into consideration in the determination of the tax rate, such as adjusting the tax rate in accordance with both the quantity and the quality of the information collected and taking into account the sensitivity of the information.[142]

Ben-Shahar asserts that the most appropriate taxpayers are data givers, meaning users, and not data takers, meaning data aggregators. He emphasises that a data pollution tax should ideally represent a normative shift, by reinforcing the idea that the main issue arising out of data pollution is not a threat to people's privacy but an accentuated need of protecting the public ecosystem. According to the contributor, users giving away their data are not those in need of protection, but those from whom the ecosystem has to be protected. The argument goes on to state that users often do not realise that they are paying for digital services with data, and this lack of awareness can be mitigated by means of the data pollution tax, which would alert users to the implicit data sharing choice that they are making.[143]

Ben-Shahar concludes by stating that data pollution tax is an idea at its embryonic stage and not a full-fledged proposal ready for implementation. The contributor underlines that practical problems are profound.

Implementation of a Pigouvian tax was one of the options considered by the authors of this book during the initial stage of their research. It is undisputed that some users do share too much data at the expense of the public ecosystem and that "failure of some members of the network to match the precaution level undermines the efforts by others".[144] That being said, establishing a Pigouvian tax seems counterintuitive for a number of reasons. First, independent from the fact that whether they are willingly sharing their data or not, users undisputedly occupy one of the most vulnerable positions in the data market. Regardless of the designated taxpayer, data aggregators or users, the burden of the tax, either in its entirety or partially, will inevitably fall on users. As a consequence, the tax will constitute an additional burden for the most vulnerable party in the data market. Second, users may also generate data in a passive manner, without realising that they are sharing their data with a data aggregator and the extent of such data. In the absence of a transaction between data aggregators and users, such data share cannot be submitted to tax, which will produce a distortionary effect on the market. Third, a small amount of Pigouvian tax levied at the time of transaction may not reveal to be very efficient in preventing users from sharing their data, especially if the tax is not levied in a salient manner.[145] Last but not least, implementation of a Pigouvian tax in the context of online transactions may reveal to be politically challenging, since the

[141] Ben-Shahar (2019), pp. 139–140.

[142] Ben-Shahar (2019), pp. 140–141.

[143] Ben-Shahar (2019), pp. 141–142.

[144] Ben-Shahar (2019), p. 116.

[145] Chetty et al. (2009).

majority of users, which is likely to constitute a non-negligible part of the population of a country, risk to react in a negative manner to such a tax.

5.2.2 Royalty on User-Generated Value

Policy makers could try to make a case in favour of taxing the value generated by users under the corporate income tax. With this in mind, Aslam and Shah analysed—among other alternatives, such as an extended PE concept—a potential *royalty on user-generated value*.[146]

Their proposal is based on an analogy between the collection of user data and the extraction of natural resources. While the analogy with natural resources is not exact (data are not rival), there are non-negligible similarities between data and natural resources: both are excludable, both might be seen as collective national assets and both need to be refined and processed.[147] More importantly both are unique to a particular location.[148] Data are, indeed, about users in particular locations.[149] Consequently, just as the rents associated with natural resources are seen as a proper object of taxation where they are located, so might be the rents associated with data relating to a country's residents.[150]

The purpose followed by tax regimes in relation to extractive industry is "to maximize compensation to host governments, while maintaining the investor's incentives to undertake exploration and extraction, given their respective risk profiles".[151] The idea put forward by Aslam and Shah is to design a compensation regime ensuring a fair level of government revenue, combined with a sufficient return to the investor. Under their view, the country involved should receive a compensation for extraction of or access to the national data asset.[152] The contributors underline that such compensation may be obtained via a number of alternative tax instruments, but they opt for focusing on the design of a royalty on data collection.[153]

In this context, royalties can be defined as payments that companies must make to governments for permission to engage in certain activities.[154] Royalties, typically used in relation to extractive industries, tend to complement corporate income

[146] Aslam and Shah (2020).

[147] IMF (2019), p. 16. For more information on this point, see also Sect. 2.3.

[148] IMF (2019), p. 16.

[149] Cui (2019), (electronic version).

[150] IMF (2019), p. 16.

[151] Aslam and Shah (2020), p. 37.

[152] Aslam and Shah (2020), p. 37.

[153] Aslam and Shah (2020), p. 37.

[154] Aslam and Shah (2020), p. 37.

taxes.[155] They form a useful instrument—albeit an imperfect one—in cases where direct taxation of rents is difficult, particularly for sectors heavily reliant on hard-to-value assets.[156]

The contributors consider different options for the design of the royalty base. The ideal design would consist in a simple *ad-valorem* royalty based on the value of the flow of data collected from the country and used by a company in a given year.[157] If the firm is the owner of the data, the stock of data held could also be taxed on a periodic basis.[158] Since these two options require standardised data valuation methods and/or the establishment of a spot market of data, both of which are currently lacking, Aslam and Shah consider an alternative two-step approximation for user-generated value.

The first step consists of an apportionment of a company's final user-generated revenues between countries.[159] It necessitates the determination of a method for apportioning user-generated revenues by location to approximate the relative value of the user base in different countries.[160] Establishment of such a method is likely to reveal to be challenging due to the inherent characteristics of data.[161] In addition, if a multilateral consensus is not reached with regard to the applicable method, multiple inconsistent approaches risk to arise.[162]

The second step consists of allowing for a standardised netback deduction from the country-specific revenues, to account for value added from post-data collection processing and analysis.[163] The netback deduction could be defined on a company-by-company basis—given different data refinement processes and cost structures—or in legislation at broader industry levels to ease administration.[164] Based on the assumption that MNEs are collecting data of similar quality from users worldwide and centralising their processing operations, which implies a common cost structure, it could be argued that this netback should be standardised across countries.[165]

These two steps would provide for a proxy for the value of a country's user data, which can then serve as the base of the royalty.[166] The contributors argue that rates to be applied on that royalty base should be set at an appropriately modest level, as they are applied to revenues rather than profits.[167] The countries with smaller consumer markets, due to, among others, fewer active users or lower purchasing power, may

[155] Aslam and Shah (2020), p. 38.
[156] Aslam and Shah (2020), p. 38.
[157] Aslam and Shah (2020), p. 38.
[158] Aslam and Shah (2020), p. 38.
[159] Aslam and Shah (2020), p. 39.
[160] Aslam and Shah (2020), p. 39.
[161] Aslam and Shah (2020), p. 39.
[162] Aslam and Shah (2020), p. 39.
[163] Aslam and Shah (2020), p. 39.
[164] Aslam and Shah (2020), p. 39.
[165] Aslam and Shah (2020), pp. 39–40.
[166] Aslam and Shah (2020), p. 39.
[167] Aslam and Shah (2020), p. 40.

choose to set rates lower so as not to disincentivise data extraction by non-resident digitalised businesses or deter market entry altogether.[168] In setting the rate, policymakers should also take particular care to avoid the risks of excessive pass-through to consumers.[169]

In our opinion, the data-natural resource analogy is appropriate. Data collected from users of one country can be used to analyse the common trends, dominant cultural components, main habits and other similar local traits and to produce goods and services on the basis of these local characteristics. Such an analysis may also reveal detailed information on the residents of a country, to the point where, the Chinese government considers the substantial amount of user data stored by different platforms as a 'serious threat to national security' if leaked or exposed in unwanted manners.[170] Data collected in one country are predominantly used for the purposes of producing services and products relating to the residents of the country concerned. For example, data collected from users located in Switzerland are usually not employed to render services in Australia. From this perspective, data indeed form *a national asset.*

It is important to note that Marian seems to agree with the data-natural resource analogy as well. The commentator expresses his point of view as follows: "You collect the data where the data [are]. People live, work and consume somewhere. This somewhere is where data [are] collected from, and transmitted to. This is similar to taxation of mineral exploration. *Activity must happen where the minerals are*".[171] The contributor illustrates this point by the example of Google. In the context of the data-flow tax that he suggests;[172] where Google is a resident, what it does with a specific user's data, where it analyses them, or when and how it monetises them, are not relevant. If Google collects a user's data when the user employs its search engine in the US, Google should pay the tax in the US. If a user, who is normally a resident of the US, travels to Spain and Google sends that user a push notification to try to make the user buy something in that latter country, the tax should be paid in Spain. The activity takes place where the valuable resources are, so should the taxation.[173]

In our view, a royalty on user-generated value may constitute an efficient solution. The problem is such a solution is rather complex to implement. The necessity of an international collaboration and a multilateral consensus adds another layer to this complexity. Some countries may simply not agree with the argument that the tax treatment of data should be akin to natural resources. The most obvious counterarguments to the data-natural resource analogy are expressed by Goulder in one paragraph, which reads as follows: "Let's assume we accept the premise that individual users aren't being adequately compensated for the gathering of the data

[168] Aslam and Shah (2020), p. 40.

[169] Aslam and Shah (2020), p. 40.

[170] Ben-Shahar (2019), p. 114.

[171] Marian (2022), p. 564. Emphasis added by the authors.

[172] See the following Sect. 5.2.3.

[173] Marian (2022), p. 564.

they generate. Some type of appropriation is occurring, but does it necessarily follow that the state is justified to act on its residents' behalf by demanding a 'price' [...]? Are collective internet clicks really the same kind of resource as fossil fuels lying deep underground, or lumber contained in forests? The bounties of the land and earth were created by a natural process that occurred with the passage of time and no human involvement. In contrast, our digital data exist [...] only because private industry has provided us with the technology to do so. Since governments aren't regulating digital sectors as utilities, like electricity or water, it seems a stretch to grant them the ability to recoup compensation on our behalf when our data [are] being gathered".[174]

Goulder also underlines that a depletion does not occur in the context of data, which can be used again and again. According to this contributor, data are more akin to renewable resources, like solar or wind energy; and, due to this very reason, the case for severance-type taxes[175] seems to break down.[176] While the authors of this book agree with the observation that data are indeed more akin to renewable resources,[177] this resemblance does not necessarily imply that data cannot be submitted to a tax regime that is similar to the one employed for natural resources. As indicated by Aslam and Shah as well as the IMF, the data-natural resource analogy is not an exact one.[178] Indeed, Aslam and Shah expressly underline the non-rival nature of data.[179] The analogy is drawn on the basis of the similarities between data and natural resources, as well as the appropriateness of using a similar tax regime. That being said, the arguments put forward by Goulder is likely to be advanced by the opponents of the implementation of a royalty on user-generated value and form a significant obstacle in reaching an international consensus.

5.2.3 Data-Flow Tax

Arguing that data may form a better proxy than money for the ability to pay and that source and residence no longer provide a theoretical underpinning to identify who gets tax value and where, Marian suggests the implementation of a data-flow tax.[180] The proposed data-flow tax is based on three principles: (1) the taxable base should be the volume and not the value of raw data; (2) all uses of data should be included in the tax base; and (3) the taxpayer should be the user of the data.[181]

---

[174] Goulder (2020), pp. 560–561.

[175] Severance taxes are taxes levied on the extraction of natural resources.

[176] Goulder (2020), p. 561.

[177] As expressed in Sect. 2.3, entitled "Data Are 'the Oil on Steroids'".

[178] Aslam and Shah (2020), p. 12; IMF (2019), p. 16.

[179] Aslam and Shah (2020), p. 12.

[180] Marian (2022), p. 560.

[181] Marian (2022), pp. 560–561.

The contributor establishes the *tax base* as the volume of raw data, as opposed to the monetary value of data. Under the data-flow tax, the tax base is thus measured in gigabytes, not in dollar amounts. The determined tax base has the merit of being self-adjusting: the more data are used, the higher the tax.[182]

All uses of data are included in the tax base, which eliminates the need to pinpointing the location where the data are analysed and 'value is created'. The tax is, therefore, collected on the flow of data, meaning on both upload and download of data.[183] Marian underlines that the flow of data is easy to measure and localise, given the fact that data, at the end of the day, require a physical infrastructure to be uploaded from, downloaded to, or simply to flow.[184]

Under the proposed framework, *taxpayers* are the users of data, regardless of how much data they are using and for what purpose. Every person that uploads data to the Internet or downloads data from the Internet is a taxpayer.[185] Despite the fact that this definition would normally give rise to a substantially large pool of taxpayers, the design of the tax includes generous exemptions, which would have the effect of narrowing the number of taxpayers falling within the scope of the data-flow tax to a great extent. The tax will be imposed only after a certain large volume of data has been transmitted within the taxable period, which will have the effect of capturing exclusively 'heavy users', for whom big data collection and analysis is an integral part of the business model.[186]

Marian argues that data-flow tax will improve the progressivity of the tax system. Currently, data-rich taxpayers are able to avoid income tax, as it is well documented, while low and middle-income 'data-poor' taxpayers, which are likely to rely more heavily on brick-and-mortar operations, cannot. A tax levied on data will predominantly burden the taxpayers who are able to avoid income taxation due to their shift to data-reliant business models, restoring thereby progressivity in the system.[187]

According to the contributor, the suggested data tax is not likely to produce a major elastic response in the collection of data. So long as the tax does not become prohibitively expensive and an opportunity to make profits exists, data collection will pursue at its usual locations. Data aggregators, which use data in their commercial activities, will not stop collecting data just because it became more expensive to do so. A data-flow tax, which is not extremely high to the point of being confiscatory, will only produce an effect of causing data aggregators to share a larger part of their wealth with governments around the world.[188] A data-flow tax may potentially induce service providers to pass the burden to users by charging users with a monetary fee for the use of services that are currently subject to the digital barter.

[182] Marian (2022), p. 561.
[183] Marian (2022), pp. 561–562.
[184] Marian (2022), p. 562.
[185] Marian (2022), p. 562.
[186] Marian (2022), p. 562.
[187] Marian (2022), p. 563.
[188] Marian (2022), p. 564.

Marian is of the opinion that this is not necessarily a negative outcome. The contributor underlines that if the price of a service increases (or, in the case at hand, if a service passes from 'apparently free' to a chargeable service), the number of individuals using the service is likely to decrease, which may be beneficial for the society as a whole in the context of certain digital services, like social media. In addition, if companies that are currently entering into digital barters with users opt for putting a monetary value on the collection of information of residents of a particular jurisdiction, they can be submitted to tax on the basis of the traditional methods of taxation, which would reveal to be even more beneficial for governments.[189]

Marian also underlines that the data-flow tax can produce the intended functionality of a Pigouvian tax, since it will target only taxpayers who use data as an integral part of their business model, meaning, the ones causing the negative externality.[190]

In addition, the contributor argues that a data-flow tax is easy to administer, at least when compared to income tax. Measuring the volume of data flowing through their physical Internet infrastructure should not constitute a problem for governments. Identifying the sender (uploader) and the recipient (downloader) of data also constitutes a rather uncomplicated task with the available technologies. An IP address or a physical machine address suffices. As a matter of fact, the data-flow tax is intended to be imposed upon the sender or the recipient, regardless of who, at the end of the day, is the 'beneficial owner of value'. Accordingly, the owner of the IP address or the machine is considered to be the taxpayer. If Virtual Private Networks (VPNs) are employed, the payor of the tax would be the VPN service provider. Last but not least, since governments control, or regulate, tangible Internet infrastructure, which is needed for the transmission of data, they are in position to enforce tax on data.[191]

Finally, the commentator emphasises that the implementation of a data-flow tax does not necessitate an international coordination *per se*. The tax can be designed as a purely domestic tax. This is due to the fact that no double taxation can occur under a data-flow tax framework. The purpose of the tax is to measure the data use and tax the transaction; i.e., both at an uploader and at a downloader level, given the fact that upload and download constitute two different uses of data.[192]

In our opinion, the data-flow tax would indeed form a viable solution to the extent where its personal scope of application is indeed limited to 'heavy data users'. Levying the tax at the level of the regular user, regardless of how insignificant the amount may be, would not constitute the best option. From an ability to pay perspective, it seems more appropriate to 'exclusively' burden data aggregators, at least by design.[193] It may also be argued that the data flow-tax may raise consequential

[189] Marian (2022), pp. 564–565.

[190] Marian (2022), p. 566.

[191] Marian (2022), pp. 566–567.

[192] Marian (2022), p. 568.

[193] The burden of a tax is generally shared among participants in a market (tax incidence) and does not fall in its entirety on the designated taxpayer.

issues with regard to data protection. As a matter of fact, the implementation of the tax requires a rather close monitoring of the senders and the recipients of data by governments, and may also give rise to a monitoring effectuated by some service providers. That being said, the majority of, if not all, 'data taxes' would inevitably suffer from a data protection problem. Technological advances combined with an appropriate legal system may help mitigate such issues in a significant manner.

5.2.4 Data Dependence Tax to Fund a Data Dividend

'Data dividend' is a suggestion that has been made by a number of commentators; Facebook co-founder Chris Hughes,[194] American politician Andrew Yang[195] and California Governor Gavin Newsom,[196] to name a few. The concept of 'data dividend' does not refer to a new type of tax but to a specific allocation of funds collected from data-rich companies by means of a tax. The basic idea is summarised by Hughes in the following manner: " [. . .] create a new tax on the revenues of any companies that collect and store meaningful amounts of information and data about people to build their businesses. The revenue from this data tax would fund a data dividend to every American each year. Once a year to start—and potentially monthly later on—every American would receive a check as a royalty payment for the data they create".[197]

Hughes takes the Alaskan Permanent Fund as a template to detail his suggestion. The Permanent Fund is a savings account that was created in the 1970s, at the moment of the oil boom. A portion of the gross revenues of oil companies that is paid to the State is earmarked to fund a savings account for the people. The 2.5% of the account's total value is distributed to the residents of Alaska each year. The distribution is effectuated evenly between every resident of Alaska, which amounts to about USD 1500 per year per person.[198]

In Hughes' view to form a fund that will subsidise the data dividend, any data-rich large company (not only GAFA but also any other companies that are making use of data in a substantial manner, such as banks, insurance companies and large retail outlets) may be subject to a 'data tax' of 5% on their gross revenues. The commentator argues that such a tax could raise around USD 100 billion a year, which could translate into a USD 400 yearly-check for all American adults. In this regard, Hughes underlines that a compensation for teenagers may also be considered given the fact that they tend to use the Internet more heavily. Hughes estimates that the fund would grow very large very fast, since the amount of data that are being produced and the

---

[194] Hughes (2018).
[195] Rinehart (2020).
[196] The Data Dividends Initiative (2022).
[197] Hughes (2018).
[198] Hughes (2018).

profits deriving from them would grow over the years. The commentator expects that the data dividend may allow each individual to receive well over USD 1000 in the next decade.[199]

Feygin et al. (The California Data Dividend Working Group) argue that state spending on public services and baby bonds (or universal saving accounts) constitute distribution methods that would reveal to be more efficient than *per capita* payments. According to the contributors, using the 'data dividend' to improve public services (especially healthcare and education) and to seed 'baby bonds'—i.e., guaranteed saving accounts at birth—for all residents would be more effective in reducing inequality and its structural causes.[200] In addition, using the data dividend to create public goods and services better reflects the fact that the underlying value of data comes from aggregation rather than individual inputs.[201]

In our view, using the data dividend to support the provision of public services would, indeed, constitute a more appropriate solution. A part of the fund can also be employed to subsidise unemployment related to technological advances.[202]

Clearly, distribution of a data dividend necessitates the existence of a fund created to this effect. The key question is how to constitute such a fund. Commentators supporting data dividend suggest that a new tax should be implemented to that end. Some legislative proposals in this vein were already made. For instance, Norma Smith, a Washington representative at date, proposed in March 2017, the addition of a 3.3% business-and-occupation tax on gross receipts from the sale of personal data relating to Washington State residents.[203] An even more recent example are the two bills proposed in New York in 2019, which suggest to add an additional 5% tax on the gross income of corporations that derive income from the data that residents of New York share with corporations.[204] None of the bills were adopted at the time of preparation of this manuscript.

Feygin et al. prepared an elaborate report on data dividend, which also includes suggestions for new taxes that may be implemented to raise the necessary funds. The commentators recommend a 'data dependence tax' based on the calculation of firm data exploitation and sales apportioned revenue, either alternatively or (and preferably) combined with a 'data sales tax', which consists of a tax on direct sales of consumer data to third parties.[205]

Under the 'data dependence tax', would be submitted to tax the sales apportioned revenue of firms that use large amounts of personal data. According to the commentators, while the tax can be implemented either as a marginal user tax or as a flat based tax on data intensity, the marginal user tax constitutes a superior policy

[199] Hughes (2018).

[200] Feygin et al. (2021), p. 16.

[201] Feygin et al. (2021), p. 16.

[202] For more details on unemployment related to technological advances, see also Chap. 7.

[203] Marian (2022), p. 570.

[204] Marian (2022), p. 570.

[205] Feygin et al. (2021), p. 13.

instrument.[206] Implementation as a marginal user tax would be effectuated in the following manner: (1) several tax brackets that depend on the number of (California) users whose data are stored by a firm would be created; and (2) the tax brackets would then be assessed against the firm's revenue.[207] Imposition of a flat tax can be organised in the following manner: (1) 'yes or no' criteria for assessing data intensity are determined; and (2) all firms meeting 'the yes threshold' are submitted to a flat tax.[208] Feygin et al. underline that the taxable base should consist of revenue and not profits, due to the fact that in the current economy, data-dependent firms often have low profitability but high revenues.[209] As per the suggestion made by the commentators, sales apportionment of revenues can be achieved through a combination of two assessments: percentage of revenues generated in a given jurisdiction (for the purposes of the proposal the State of California) and percentage of users domiciled in the jurisdiction concerned (again, the suggestion is made for California).[210] The second tax suggested by the commentators, namely the 'data sales tax', consists of a flat rate tax levied on the sale of data to data brokers.[211]

Compensating users for the raw data they are producing is of the utmost importance, both from a national and from an international perspective. In our opinion, however, such compensation should not unnecessarily distort the market and would, thus, best be achieved by the establishment of an efficient data labour market combined with pay-for-privacy options.[212] Since such a solution cannot be implemented at this stage in time, the second-best option appears to be the implementation of a 'data collection tax', examined in the following chapter. Our proposal diverges from the proposals made by Feygin et al., since it does not rely on an apportionment of revenues or data transactions between data aggregators.

References

Alvarado M (2021) Digital services taxes across Europe in the midst and aftermath of the Covid-19 pandemic: a plausible option to raise tax revenue? Eur Tax IBFD 61(9):403–410

Aslam A, Shah A (2020) Tec(h)tonic shifts: taxing the "digital economy". IMF staff discussion note, IMF

Avi-Yonah RS, Kim YRC, Sam K (2022) A new framework for digital taxation. Harv Int Law J 63(2):279–341

Bauer M (2018) Five questions about the digital services tax to Pierre Moscovici. ECIPE occasional paper, no. 04/2018, European Centre for International Political Economy (ECIPE)

[206] Feygin et al. (2021), pp. 14–15.

[207] Feygin et al. (2021), p. 14.

[208] Feygin et al. (2021), p. 14.

[209] Feygin et al. (2021), p. 15.

[210] Feygin et al. (2021), p. 15.

[211] Feygin et al. (2021), p. 15.

[212] As examined in Chap. 4.

Becker J, Englisch J (2019) Taxing where value is created: what's 'user involvement' got to do with it? Intertax 47(2):161–171

Becker J, Englisch J, Schanz D (2019) Re-allocation of taxing rights for big data business models. https://doi.org/10.2139/ssrn.3433715. Accessed 07 July 2023

Ben-Shahar O (2019) Data pollution. J Leg Anal 11:104–159

Brauner Y (2022) Agreement? What agreement? The 8 October 2021, OECD statement in perspective. Intertax 50(1):2–4

Bruins G, Einaudi M, Seligman E et al (1923) Report on double taxation, League of Nations Economic and Financial Commission, Document E.F.S.73.F.19

Chetty R, Looney A, Kroft K (2009) Salience and taxation: theory and evidence. Am Econ Rev 99(4):1145–1177

Collin P, Colin N (2013) Task force on taxation of the digital economy, Report to the Minister for the Economy and Finance, the Minister for Industrial Recovery, the Minister Delegate for the Budget and the Minister Delegate for Small and Medium-Sized Enterprises, Innovation and the Digital Economy

Cui W (2019) The digital services tax: a conceptual defense. Tax Law Rev 73(1):69–11. (electronic version)

Cui W, Hashimzade N (2019) The digital services tax as a tax on location-specific rent, CESifo working paper, no. 7737. https://www.econstor.eu/bitstream/10419/201963/1/cesifo1_wp7737.pdf. Accessed 07 July 2023.

Das RR (2020) The concept of value creation: is it relevant for the allocation of taxing rights. Bull Int Tax IBFD 74(3):134–143

ECOFIN (2019) ECOFIN Report to the European Council on tax issues – Endorsement, Doc. 9773/19 FISC 281 ECOFIN 528

Elliffe C (2021) Taxing the digital economy: theory, policy and practice. Cambridge Tax Law Series, Cambridge University Press

European Commission (2018a) COM (2018) 147 final 2018/0072 (CNS), Proposal for a Council Directive laying down rules relating to the corporate taxation of a significant digital presence

European Commission (2018b) COM (2018) 148 final 2018/0073 (CNS), Proposal for a Council Directive on the common system of a digital services tax on revenues resulting from the provision of certain digital services

Feygin Y, Hecht B, Prewitt M et al (2021) A data dividend that works: steps toward building an equitable data economy, Berggruen Institute. https://www.berggruen.org/themencode-pdf-viewer-sc/?tnc_pvfw=ZmlsZT1odHRwczovL3d3dy5iZXJnZ3J1ZW4ub3JnL3dwLWNvbnRlbnQvdXBsb2Fkcy8yMDIxLzA1L0RhdGEtRG12aWRlbmRfZmluYWwucGRmJnNldHRpbmdzPTExMTExMTExMTExMTExMTExJmxhbmc9ZW4tVVM=#page=&zoom=auto&pagemode=. Accessed 07 July 2023

Goulder R (2020) Should data extraction be taxed as a natural resource? Tax Notes International, 27 July 2020, pp 559–561

Herzfeld M (2017) The case against BEPS - lessons for coordination. University of Florida Levin College of law research paper no. 18-3

Hey J (2018) "Taxation where value is created" and the OECD/G20 base erosion and profit shifting initiative. Bull Int Tax IBFD 72(4/5):203–208

Hughes C (2018) The wealth of our collective data should belong to all of us. The Guardian. https://www.theguardian.com/commentisfree/2018/apr/27/chris-hughes-facebook-google-data-tax-regulation. Accessed 5 Sept 2022

IMF (2019) Corporate taxation in the global economy. IMF policy paper no. 2019/007, IMF

Jiménez AM (2018) BEPS, the digital(ized) economy and the taxation of services and royalties. Intertax 46(8/9):620–638

Kofler G, Sinnig J (2019) Equalization taxes and the EU's 'digital services tax'. In: Haslehner W, Kofler G, Pantazatou K, Rust A (eds) Tax and the digital economy: challenges and proposals for reform. Wolters Kluwer, pp 101–146

Lammers J (2020) Less is more - can developing countries gain tax revenue by giving up taxing rights? Copenhagen Business School Research paper no. 20-16

Laudon KC (1996) Markets and privacy. Commun ACM 39(9):92–104
Marian O (2022) Taxing data. Brigham Young Univ Law Rev 47(2):511–576
OECD (2018) Tax challenges arising from digitalisation – interim report 2018: inclusive framework on BEPS, OECD/G20 Base Erosion and Profit Shifting Project, OECD Publishing
OECD (2019) Public consultation document, addressing the tax challenges of the digitalisation of the economy, 13 February - 6 March 2019. OECD Publishing
OECD (2020a) Tax challenges arising from digitalisation – economic impact assessment: inclusive framework on BEPS, OECD/G20 Base Erosion and Profit Shifting Project, OECD Publishing
OECD (2020b) Tax challenges arising from digitalisation – report on pillar one blueprint: inclusive framework on BEPS, OECD/G20 Base Erosion and Profit Shifting Project, OECD Publishing
OECD (2021) Statement on a two-pillar solution to address the tax challenges arising from the digitalisation of the economy, 8 October 2021, OECD/G20 Base Erosion and Profit Shifting Project, OECD Publishing
OECD (2022a) Two pillar solution to address the tax challenges arising from the digitalisation of the economy, frequently asked questions, July 2022. https://www.oecd.org/ctp/beps/faqs-two-pillar-solution-to-address-the-tax-challenges-arising-from-the-digitalisation-of-the-economy-july-2022.pdf. Accessed 07 July 2023
OECD (2022b) Public consultation document, pillar one – amount a: draft model rules for domestic legislation on scope, 4 April 2022 - 20 April 2022. OECD Publishing
OECD (2023) Outcome statement on the two-pillar solution to address the tax challenges arising from the digitalisation of the economy, 11 July 2023. https://www.oecd.org/tax/beps/outcome-statement-on-the-two-pillar-solution-to-address-the-tax-challenges-arising-from-the-digitalisation-of-the-economy-july-2023.pdf. Accessed 12 July 2023
Parsons A (2022) Tax's digital labor dilemma. Duke Law J 71:1781–1847
Pigou AC (1932) The economics of welfare, 4th edn. Macmillan and Co. Ltd.
Pistone P, Ullmann A (2021) Digital taxes and Article 2 OECD Model Convention 2017. In: Kofler G, Lang M, Pistone P et al (eds) Taxes covered under Article 2 of the OECD model: the scope of tax treaties in a dynamic global environment of newly created taxes, 1st edn. IBFD, pp 167–199
Postler (2020) The OECD's work on profit allocation and nexus rules for a digitalized economy – a potential improvement of the international taxation framework? Bull Int Tax IBFD 74(2):76–87
Rinehart W (2020) Andrew Yang's plan to pay you for your data doesn't add up. Wired. https://www.wired.com/story/opinion-andrew-yangs-plan-to-pay-you-for-your-data-doesnt-add-up/. Accessed 07 July 2023
Schön W (2018) Ten questions about why and how to tax the digitalized economy. Bull Int Tax IBFD 72(4/5):278–292
Schön W (2019) One answer to why and how to tax the digitalized economy. Max Planck Institute for Tax Law and Public Finance Working Paper 2019-10, June 2019
The Data Dividends Initiative (2022) Homepage. https://www.datadividends.org. Accessed 07 July 2023

Chapter 6
Ending the Era of 'Free' Data: Data Collection Tax (Step 1)

Abstract While waiting for an efficient data labour market to be established, an interim measure reveals to be necessary to safeguard the interests of users. As this 'second best' solution, the authors suggest the implementation of a 'data collection tax'. This chapter begins by elucidating the rationale behind this new type of tax and the main idea. It proceeds with the examination of the personal scope of application of the tax, the material scope of application of the tax, the taxable base and the tax rate along with the other relevant features of the tax, namely the need for a robust data protection legislation and system, implementation of a right to opt-out from generation of data and deductibility of the amount of data collection tax from the taxable base of the corporate income tax. Then, are briefly described the consequences of the data collection tax (i.e., data minimisation and rupture of the digital barter) and potentially controversial issues (i.e., pass-through of the costs onto users and discrimination of certain highly digitalised businesses). The chapter concludes by making recommendations on the allocation of the funds raised via the data collection tax.

The data collection tax that the authors suggest implementing is positioned in *step one*, which regroups operations related to data generation by users and data collection by data aggregators. The data collection tax is thus levied in generation and collection phases of the data's life cycle.[1]

The authors envision the data collection tax as an *interim measure*. As concluded in Chap. 4, the establishment of a data labour market with an option to pay-for-privacy for users[2] who do not wish to take part in the market concerned, would constitute the 'best' solution for the two key problems occurring in *step one*, namely the lack of privacy and the lack of adequate compensation for user data. Such a market would benefit to all its participants. Users would be able to protect their

[1]For more details on the four steps of the data market, see Sect. 3.5.

[2]To guarantee the ease of reading, for the purposes of this chapter, the reference is only made to a 'data labour market' for referring to a data labour market combined with an option to pay-for-privacy for users.

X. Oberson, A. E. Yazıcıoğlu, *Taxation of Big Data*,
https://doi.org/10.1007/978-3-031-43372-6_6

privacy and get a fair compensation for the data they opt for sharing. Data aggregators would be able to collect data in a more apparent and organised manner, since they would begin to purchase this new raw material directly form 'its source' in exchange for an equitable payment. Such an open data collection process may eliminate information mismatches currently occurring between the data collected and the data actually needed by the aggregator. A market entails, by its very design, the possibility to specifically target certain data, by overtly requesting the data needed from users. A more effective collection process would also reduce the ROT data. With proper legislation and governmental supervision, a data labour market could function as any other market and self-regulate to a great extent. Organising data transactions through an efficient and largely self-regulating market would have the merit of minimising any economic distortions that may occur following the 'newly established' necessity to adequately compensate user-generated data.

Instituting a data labour market is, however, currently not possible for a variety of reasons including technical limitations.[3] This impediment does not justify and should not result in a wait-and-see situation, since there is a clear and ever-increasing need to safeguard the interests of the most vulnerable actor of the data market: users. A 'second best' solution needs to be identified and implemented, albeit as a temporary measure. The authors consider that this second-best solution is a 'data collection tax'.

This chapter examines the main features of the suggested data collection tax. The purpose is not to present a tax that is ready to implement, but to provide for the primary characteristics that such a tax should have. A thorough analysis of a number of aspects, including the determination of an approximate value of data per user, would be necessary to effectively bring the data collection tax into life. That being said, the tax remains an easy-to-implement and efficient option, which would cause only minimal economic distortions. Moreover, notwithstanding its classification as a second-best option, on a variety of points, the data collection tax gives rise to certain advantages that cannot be obtained through the establishment of a data labour market. Such advantages are also underlined throughout the analysis conducted below.

First, the idea is briefly described. *Second*, the taxpayers are identified and tax-exempt entities are listed. *Third*, the material scope of application is determined. *Fourth*, the taxable base and the tax rate are established. *Fifth*, the other relevant features of the tax are detailed. *Sixth*, the actual consequences of the tax, from both a legal and an economic perspective, along with issues that may be subject to debate are enumerated. *Seventh*, suggestions are made for the allocation of the funds collected from the data collection tax.

[3]For more information on this point, see Chap. 4.

6.1 Idea

The main purpose of the data collection tax is to charge data aggregators for the 'raw material'[4] that they are currently collecting either for free or in exchange for digital services. While under a data labour market, the payment for a single datum, or a collection of 'datum', obtained from a user is directly made to the user concerned; under the data collection tax, the payment for the collective data obtained from users residing in one country is made to the government of the country concerned.

In a nutshell, the rationale behind the data collection tax consists of implementing a tax that is closely in touch with the economic and factual reality, while not giving rise to any legal controversies. The tax is intended to form, as much as possible, an organic part of the data market.

Four aspects need to be developed in detail to express the main idea: rationale behind designating the government as a proxy, choice of users' residence as a criterion, relevance of collective data and requirement of payment in exchange for raw material.

The data collection tax suggested by the authors demonstrates certain similarities to the 'Excise Tax on the Collection of Consumer Data by Commercial Data Collectors'[5] of the State of New York (NY Excise Tax), currently in Committee,[6] introduced by Senator Krueger.[7] Accordingly, reference is also made to certain articles of the NY Excise Tax Bill, to the extent relevant for the purposes of the elucidation of the data collection tax proposal put forward by the authors.

6.1.1 Rationale Behind Designating the Government as a Proxy

The idea of a data collection tax draws, to a certain extent, on the data-natural resource analogy. Users are, just like natural resource assets, immobile at and unique to a particular location.[8] The data being collected by different data aggregators are specifically about these users in distinct locations.[9] As a matter of fact, data collected from users of one country are employed in view of analysing the common trends, dominant cultural components, main habits and other similar local traits, as well as

[4] On this point, see Sect. 2.3.

[5] Senate Bill S2012, Assembly Bill A3959. Hereinafter, 'the NY Excise Tax Bill'.

[6] The progress can be tracked via https://www.nysenate.gov/legislation/bills/2021/s4959. Accessed on 07 July 2023.

[7] It is important to mention that Avi-Yonah et al. suggested a similar tax, which diverges, however, to a great extent from the tax being suggested by the authors [on this point, see Avi-Yonah et al. (2022), pp. 335–340].

[8] IMF (2019), p. 16.

[9] Cui (2019), (electronic version).

of producing goods and services on the basis of these *local characteristics*. Accordingly, data collected in one country are predominantly used for the purposes of producing services and products relating to the residents of the country concerned. For example, data collected from users located in Ireland are usually not employed to render services in France. Also, analysing substantial amounts of data belonging to the residents of one country provides detailed sensitive information on the residents of the country concerned, such as their political preferences and the most frequent health problems. On this basis, it can be stated with ease that data indeed form *a collective national asset*.[10]

In the case of natural resources, which also form national assets, governments typically hold the resources in trust on behalf of their citizens and collect payment for their extraction in the form of government revenues.[11] Just as the rents associated with natural resources are seen as a proper object of taxation where they are located, so should be the rents associated with information about a country's residents, under the current circumstances.[12] Monetisation of this collective national asset may, therefore, represent a form of location specific rent,[13] and be submitted to tax by means of royalties, i.e., payments that companies must make to governments for permission to engage in certain activities.[14]

While not constituting a part of the rationale behind designating the government as a proxy *per se*, governments are better placed to exert collective power on behalf of users, who have an almost non-existent bargaining power with regard to their data.[15] This powerful side effect may allow the user base to benefit more from the compensation paid for their data at the end of the process, in conjunction with a proper allocation of funds deriving from the tax.

In a similar vein, the collection of the compensation by the government in the form of a tax may allow a more efficient control of the data-flow and, therefore, translate into a more accurate evaluation of the value of data. In fact, for the data to be properly monetised, freely available data should become a scarce commodity.[16] This entails an efficient control of every data aggregator, and not only multi-sided platforms. Even if the greatest portion of the users of a country refuse to enter into digital barters, data brokers may still continue to collect data without compensating users, and more importantly, without their knowledge. Governments controlling, or regulating, tangible Internet infrastructures of their country may ensure an efficient control of the data traffic and prevent any data collection process without the payment of the data collection tax.

[10] Same opinion with Aslam and Shah (2020), p. 23.

[11] Aslam and Shah (2020), p. 23.

[12] On this point, see also IMF (2019), p. 16.

[13] Cui (2019), (electronic version).

[14] Aslam and Shah (2020), p. 37.

[15] On this point, see also Aslam and Shah (2020), p. 23; and Sect. 4.7.

[16] Goulder (2020), p. 560.

6.1.2 Choice of Users' Residence as a Criterion

Natural resources are held in trust on behalf of the *citizens* of a country. For the purposes of data, the criterion of *residence of users*, as opposed to citizenship, is more appropriate. As emphasised by Marian, "You collect the data where the data [are]. People live, work and consume somewhere. This somewhere is where data [are] collected from [. . .]".[17] Users' residence is also employed by the NY Excise Tax, which applies to commercial data collectors aggregating consumer data on more than one million individual consumers, whose primary residence is in New York State.[18]

The case for choosing users' residence as a criterion could best be illustrated by means of an example relating to the information provided by consumers about themselves via, for instance, the use of a search engine.[19] Data collected about an American citizen who lives in New York via a search engine should be considered as being a part of the collective data of the US. When the same American citizen temporarily moves to Geneva (Switzerland) in the course of their employment and uses the same search engine through which their data are collected, the information concerned belong to the collective data of Switzerland. In fact, a great portion of data is being utilised to render local services on the basis of established local traits. 'Expats working in Geneva' is likely to constitute a relevant data segment and may even be commercialised for a higher price when compared to the data available on Swiss citizens living in Geneva.

In addition, citizenship demonstrates the existence of a mere legal bond, whereas residence manifests the presence of both a legal and a factual bond. Data are collected and employed through applications, websites, devices and other similar instruments that interact with users on a daily basis. They are valuable mainly because of their ability to reflect the daily lives, decisions, emotions and characteristic traits of users. Data are, thus, inextricably linked to facts. In consequence, the criterion of residence is the best-fit.

To avoid all ambiguity in practice, the concept of residence should be established as precisely as possible. To the extent relevant and needed, a specific concept of residence, putting more emphasis on physical presence, can be included in the data collection tax legislation.[20]

---

[17] Marian (2022), p. 564. Emphasis added by the authors.

[18] Article 2(A)(II) and (E) of the NY Excise Tax Bill.

[19] IMF (2019), p. 15.

[20] On this point, see also Plattner (2022), p. 1387.

6.1.3 Relevance of Collective Data

Every single datum has value. Nothing excludes a single datum from being traced, identified, separated and transferred or sold by a data aggregator. That being said, and quite reasonably, a collection of datum, especially a large collection of datum has substantially more value. While the value of the overall data collected from one user may amount to USD 548 per year when analysed separately,[21] the overall data collected from the user base of a country is worth, beyond any doubt, considerably more. Data's value is, as a matter of fact, exponential.[22] As righteously stated by Feygin et al., "[...] the economic value of data primarily comes from the data aggregation generated by large groups, rather than from any one individual [...] we must think in terms of 'our data' not 'my data'".[23]

Accordingly, the data collection tax may allow for a more appropriate compensation, which reflects the value of the data collected from a country more accurately than a full-grown data labour market. It is clear that in the context of a data labour market each use of data would give rise to a remuneration, in contrast to the data collection tax within the framework of which the compensation would only be due once at the time of collection. Nonetheless, in the former the value will always be calculated per user, while in the latter the value will be determined on a per country basis and reflect, therefore, its exponential character. Combined with an adequate allocation of funds, it is undisputable that the amount generated by the data collection tax would produce a more pronounced positive impact on users.

6.1.4 Requirement of Payment in Exchange for Raw Material

The data collection tax consists of, in fact, paying for the raw material. Governments would be receiving the compensation for their residents' data, on their behalf. Just as any company that has to pay for its raw materials, highly digitalised enterprises would begin paying an appropriate amount in exchange for their raw material as well. Subsequently, when calculating the amount of the tax, exclusively the value of the actual 'raw' user data should be taken into consideration, without any regard to any potential uses that could be made by the data aggregator following this initial acquisition. For instance, the same data could be more efficiently used and, thereby, become more valuable at the hands of Google than of a medium-scale data aggregator. The tax paid by both companies, should, however be identical, since the amount of the data collection tax should be equal to the value of raw data. Otherwise, the data collection tax would act as an alternative to the corporate income

[21] For a detailed explanation, see Sect. 4.5.2.

[22] See Sect. 2.3.

[23] Feygin et al. (2021), p. 5.

tax and give rise to an unequitable treatment of data aggregators, along with legal controversies.

6.2 Personal Scope of Application

Taxpayers of the data collection tax should consist of all data aggregating companies. Multi-sided platforms, data brokers and other data aggregators would be subject to tax. The tax would be applicable regardless of the country where data aggregators are established, i.e., both to resident and non-resident data aggregators.

To ensure the efficient functioning of the system and non-availability of data collected from users without compensation, only public institutions following public functions—and not commercial purposes—, typically health and education, would be exempt from the data collection tax. Such institutions would not be allowed to share the data concerned with private parties, with the notable exception of the existence of a prevalent public interest.

6.3 Material Scope of Application

Every piece of datum collected should fall within the scope of application of the tax. *No thresholds*, such as number of users or gigabytes of data, should be established. Once data have been assimilated to what they really are, namely a raw material, it becomes clear that every enterprise should pay for this raw material, just like any other raw materials. For instance, coal is not free of charge until an enterprise reaches a quota of 250 kg. The same should hold true for data. In the same vein, as a general principle, it is not relevant whether the data have been collected in *a digital or an analogue* manner. Data provided by means of a form filled-in by handwriting and incorporated into a database, again either analogue or digital, by enterprises should also fall within the scope of the tax. Supervision and control processes in place to monitor the protection of personal data may allow to keep track of such data. Similarly, the *type of the data* collected should not be taken into consideration while determining the tax base.[24] Volunteered or observed, captured or exhaust, all data form raw materials.

It could be argued that such a wide scope of application would result in an unreasonable burden on small and medium-sized enterprises (SMEs), including start-ups. This argument is not justified for four main reasons. *Primo*, SMEs and start-ups are paying an adequate compensation for all their raw materials. It is difficult to understand why specifically utilisation of data should give rise to an artificial deviation from this economic reality. *Secundo*, establishing thresholds

[24] For more information on the main types of data, see Sect. 2.2.

would distort the market. Small and medium-sized data aggregators would be able to collect the data without paying any compensation, aggregate such data with the data that they have or that they purchase from data brokers and form large databases. They may also sell the data at a cheaper price than the one being charged by the government to enterprises surpassing the threshold. In any case, such thresholds would unduly put SMEs at an advantageous position and erode the tax base. *Tertio*, distinguishing between the methods of collection would result in a similar issue; aggregators collecting data in an analogue manner would be unjustifiably put at a more favourable position. Also, the tax base would be eroded, since the data collected in an analogue manner could then be incorporated in a digital database and used in the same manner as 'digitally collected data'. *Quarto*, submitting to tax only certain types of data would, once again, distort the market. For instance, if only volunteered data are submitted to tax, data brokers, which mainly collect observed data, would obtain an unfair advantage over multi-sided platforms. Consequently, to safeguard the tax base, along with the neutrality of the tax, the net must be cast as wide as possible.

One exception could be made to this general rule. Data that have to be collected to complete a transaction (such as the address information of a consumer who makes an online purchase) may be considered as falling out of the material scope of application. A similar approach is followed by the NY Excise Tax, which excludes from its scope of application 'consumer contact information', consisting of a consumer's e-mail address, telephone number, telefax number, home address, mailing address and credit card information necessary to engage in a sales transaction.[25] From a purely theoretical perspective, such information constitutes a raw material necessary to commercialise a product or a service as well, and should, therefore, be included in the scope of application. That being said, this additional cost of raw material, like all costs, is likely to have the effect of increasing the prices of the goods or services concerned. From a policy perspective, it does not seem justified to make consumers pay an additional amount for the information that has to be shared to obtain a product or a service.[26]

6.4 Taxable Base and Tax Rate

In a nutshell, the taxable base would consist of data collected from users on a yearly-basis and the amount of tax would ideally correspond, where possible, to the value of the collective data to reflect the exponential value of data. For cases where such a

calculation of value is either not possible or not relevant, the tax may take the form of a unit charge per datum and per user (e.g., health data per user). By opting for a

[25] Article 2(A)(II) and (C) of the NY Excise Tax Bill.

[26] For more information on the possible effects of the tax on users, see Sect. 6.6.

comparable approach, the NY Excise Tax adopts a per consumer rate, progressing on the basis of the number of consumers whose data are collected.[27]

Establishing the exact value of data forms a quasi-impossible task as explained in Chap. 4.[28] Determining the value of data would, however, reveal to be more feasible in the context of the data collection tax when compared to a data labour market. In fact, the purpose is to establish the value of the collective data generated during 1 year in a country, and not to pinpoint the value of data on a per user and per transaction basis. Calculating the value of data remains more achievable even in cases where a unit charge per datum and per user is levied. Once again, the aim is not to compute the value of the data of a specific user under a determined transaction but to establish an annual fixed price per user and per type of data.

For large data aggregators, including multi-sided platforms, such as social media and e-commerce websites, taking into consideration the collective data would be more adequate. Such enterprises are indeed relying on substantial amounts as well as, in the majority of cases, all four main types[29] of data to conduct their business. The value of the collective data can only be calculated on a case-by-case basis, i.e., specifically for each enterprise, depending on a number of factors such as the number of users, the time spent by users to generate the data and the types of data that are being collected. For SMEs, however, which collect a much-limited amount and, in most cases, only one or two types of data, computing the amount of the collective data may not be a meaningful endeavour. In such cases, applying a unit charge per datum and per user would reveal to be more suitable. Different annual prices should be set on a per user basis for different types and categories of data. For example, volunteered health data would typically be more expensive than exhaust location data, since the former is more valuable than the latter.

Albeit being more manageable when compared to a full-fledged data labour market, the valuation of data in the context of the data collection tax undoubtedly presents a great challenge. However, this difficulty should be recognised as a practical issue and should not give rise to the alteration of the general principle that needs to be adopted for an efficient functioning of the data business.[30] Also, the challenge posed by the valuation of data has been mentioned over and over in the last decade to legitimise the absence of any tangible action. At some point, an effort has to be made to reach a solution, rather than concentrating on the problem.

As underlined by Collin and Colin, already in 2013, technological developments did give rise to a documentation of the value of data by the market and market observers, such as research and consulting firms, academics and international organisations.[31] Many consulting firms, like the McKinsey Global Institute and

[27] Article 3 of the NY Excise Tax Bill. It is to be noted that the Bill opts for a per month rate and does not distinguish between any different types of data.

[28] See Sect. 4.5.

[29] Volunteered and observed as well as captured and exhaust.

[30] IMF (2019), p. 16.

[31] Collin and Colin (2013), p. 45.

Bain&Company, drafted documents proposing an assessment of the economic value of big data in general.[32] A myriad of academic research has been effectuated on the value of personal data and their use by public-sector as well as private-sector organisations.[33] Researchers have also been working on measuring data and their value in statistics.[34] In addition, despite the lack of a professional market observation, there exist some databases and a considerable amount of information on data transactions.[35] Certain databases aggregate and structure operating and financial data of enterprises; a huge mass of information on data transactions are available from the press, blogs and social networking platforms; and increasingly sophisticated tools are being made available for integrating, analysing and visualising data from different databases.[36] Governments may rely on such databases, tools as well as relevant agencies and academic research centres to build up expertise in this area[37] and establish the value of data for the purposes of the data collection tax.

With regard to large data aggregators, which would require a case-by-case basis analysis, the amount of the tax could be assessed on the basis of dual declarations. In one declaration, the company would quantify the value of the data that it collects, subject to an audit by the tax authorities. In another declaration, the company would be required to commission external audits by independent third parties to quantify the value of data.[38] A designated governmental authority could also directly engage in talks with large data aggregators, either separately or collectively, to establish the value of the collective data, at least during the initial years of implementation of the data collection tax. It would be advisable to follow a similar path with regard to small and medium-sized data aggregators. A designated governmental authority could engage in talks with the representatives of small and medium-sized data aggregators while determining the annual value of different types of data.

6.5 Other Relevant Features

The three following aspects should be incorporated into the design of the data collection tax: (i) a robust data protection legislation and system, (ii) implementation of a right to opt-out from generation of data, and (iii) deductibility of the amount from the taxable base of the corporate income tax.

---

[32] Collin and Colin (2013), pp. 45–46.

[33] Collin and Colin (2013), p. 46.

[34] Collin and Colin (2013), pp. 46–47.

[35] Collin and Colin (2013), p. 118.

[36] Collin and Colin (2013), pp. 118–119.

[37] Collin and Colin (2013), p. 119.

[38] See also, Collin and Colin (2013), p. 123. The idea expressed by the contributors has been adapted to the data collection tax suggested by the authors of this book.

6.5.1 A Robust Data Protection Legislation and System

To enforce the data collection tax, governments should put in place a system detecting, or controlling, the traffic of data. Such system would necessarily entail some kind of government supervision over the data generation and collection process. Although the technical details are beyond the scope of expertise of the authors, the governmental system of data supervision should be constructed in the least intrusive manner possible. Should notably accompany this new system, among others, regular destruction of unnecessary data, mechanisms to prevent data breaches as well as data misuses and a detailed legislation on data protection.

6.5.2 Right to Opt-Out

The implementation of the data collection tax constitutes an occasion that could be seized to put into effect a right to opt-out for users who prefer to safeguard their privacy by not disclosing their data to any data aggregators.

This suggestion is based on the California Consumer Privacy Act, which grants such a right to consumers under its Section 1798.120. Under the Act, consumers have the right, at any time, to direct a business that sells consumers' personal information to third parties not to sell their personal information.[39] To enable consumers to use their right, businesses that sell consumers' personal information to third parties should inform the consumers about this fact and inform them about their right to opt-out.[40] The exercise of the right to opt-out should be rendered as accessible as possible to consumers, through a clear indication of 'Do Not Sell My Personal Information' options.[41] Consumers exercising their right to opt-out should not be subject to discrimination by the businesses concerned. Would amount to a discrimination, among others, denying goods and services to the consumer; charging different prices or rates for goods or services, including through the use of discounts or other benefits or imposing penalties; and provision of a different level or quality of goods or services.[42] The Act states, however, that nothing prohibits a business from charging a consumer a different price or rate, or from providing a different level or quality of goods or services to the consumer, if that difference is reasonably related to the value provided to the business by the consumer's data.[43]

The material scope of the right to opt-out should be larger than personal data, and englobe all kinds of data generated by users. As a matter of fact, some types of datum collected by data aggregators may not constitute personal data in the legal sense of

[39] California Consumer Privacy Act, Section 1798.120, (a).

[40] California Consumer Privacy Act, Section 1798.120, (b).

[41] California Consumer Privacy Act, Section 1798.135, (a).

[42] California Consumer Privacy Act, Section 1798.125, (a)(1).

[43] California Consumer Privacy Act, Section 1798.125, (a)(2).

the term. This would occur, for example, in cases where the datum, on its own, is anonymised and cannot be traced back to a specific individual. Users should still be able to opt-out of the collection of such data, given the fact that combination of large data sets consisting of 'non-personal data' may produce personal data. An efficient opt-out system should, thus, cover all types of user data.

It can be assumed that once data aggregators begin paying for raw data, they would not have an incentive to charge a fee (or a higher fee) to consumers not willing to share their data or to provide them with a good or a service of lower quality. A data collection tax may, thus, serve as a tool to implement a true right to opt-out for consumers, without any particular consequences. In the eventuality where the value of the data calculated by the governments stays lower than the actual value of the raw material, which can quite likely be the case, data aggregators may continue to have an incentive to charge a (higher) fee to consumers not disclosing their data or to provide a lower quality good or a service to them. It is important to remind that such a scenario can only arise for data aggregators collecting volunteered data. The ones aggregating observed data are usually not directly interacting with consumers.

6.5.3 Deductibility from the Corporate Income Tax Base

By design, the amount of the data collection tax corresponds to the value of the raw material. The system put in place aims to incorporate the purchase of data from users into the functioning of the data market, despite the fact that such price is being collected by the government. In accordance with this line of reasoning, the amount of tax should be characterised as a *cost* for the paying company for corporate income tax purposes. As a matter of fact, when a company purchases raw material (such as oil and coal), the amount paid for the material concerned is normally deductible from the corporate income tax base. The same principle should hold true for the data collection tax, which corresponds to a purchase of data.

It may be argued that deduction of the data collection tax should not be allowed due to data's non-rival nature. The 'justification' of such argument may be based on the fact that data do not perish with utilisation like other raw materials would. Quite the opposite, data can, not only be used over and over again for the same or different purposes, but also may become even more fruitful through utilisation.[44] Even though such an argument seems reasonable at first glance, it is not persuasive. To the extent necessary, the non-rival nature of data should be taken into consideration during the computation of the 'price' of this raw material, i.e., while determining the amount of tax. In fact, while certain types of data may have a long 'shelf life' and can be used on a countless number of occasions, other types may become rapidly outdated and unsuitable for use (in a sense 'perish'). Also, the fact that data become more fruitful

[44] For more information on the main characteristics of data, see Sect. 2.3.

through utilisation is inextricably linked to phases of processing and analytics, during which data cannot be considered as a 'raw material' anymore.

Deductibility ensures that data collection tax does not result in a double taxation. The amount of the data collection tax is not taken into consideration in the determination of the corporate income tax base. In practice, absence of double taxation would be more effectively ensured for resident companies. A country implementing a data collection tax can also make the relevant modifications in its corporate income tax law to enable its deductibility. To prevent international double taxation, however, reaching a common international understanding would, inevitably, reveal to be necessary. This international consensus would only consist of a common understanding of the countries to accept the data collection taxes for what they are: an adequate price paid for 'data' as a raw material. A sustained effort and collaboration should, under normal circumstances, not be needed.

Allowing the deductibility of the amount of the data collection tax from the corporate income tax base constitutes an essential feature of this former tax. Such deductibility is, in fact, the key element to treat the purchase of data in an identical manner to the purchase of other raw materials and to enable the data collection tax to fulfil its purpose. In other words, deductibility guarantees the neutrality of the tax.

6.6 Consequences of the Tax and Potentially Controversial Issues

The two actual consequences of the tax are *data minimisation* and *rupture of the digital barter*. The two issues that will potentially give rise to debate are a possible *pass-through of the costs onto users* and *discrimination of certain highly digitalised businesses*.

It can safely be assumed that once they begin paying a monetary compensation for the raw data they collect, data aggregators would opt for collecting only the data that are directly relevant and necessary to accomplish their business activities. In other words, they would put an end to the practice of aggregating all the data 'they come across', just in case that data might become useful in the future. Such *data minimisation*[45] would reduce the ROT data, along with the frequency and negative impact of data breaches.

Payment of the data collection tax will *rupture the digital barter*.[46] In addition of providing 'free' digital services, enterprises falling within the scope of application of

---

[45] 'Data minimisation' is one of the principles relating to processing of personal data enumerated in article 5 of the GDPR. It corresponds to collecting data adequate, relevant and limited to what is necessary in relation to the purposes for which they are processed [art. 5(1)(c) of the GDPR]. The data collected by data aggregators may not necessarily constitute personal data within the legal sense of the term. The term 'data minimisation' was opted for its appropriateness in the context of the analysis and not to imply the existence of personal data.

[46] For more information on the digital barter, see Sect. 3.5.1.

the tax will begin paying a monetary compensation for user data. This modification in the current functioning of the market may potentially pave the way for a pass-through of the costs onto users, which forms one of the potential issues related to the implementation of the data collection tax. The rupture of the digital barter also constitutes a relevant aspect for VAT purposes.[47]

It can be assumed that breaking off the digital barter would most probably result in enterprises requiring their users to pay for the digital services they are currently providing in exchange for data. Such an end result will be perceived as a detrimental development by the general public, especially when the general expectation of free Internet services is taken into consideration.[48] While a potential shift on the currently 'free' provision of digital services due to the data collection tax cannot be excluded altogether, the probability of its actual occurrence remains substantially low. Digital barter occurs in multi-sided platforms, within the framework of which users are subsidised in view of providing services to the customers of the other side(s). This subsidisation is rooted in one simple economic reality: without users' attention (which seems to almost amount to 'devotion' in certain cases), time and data, such platforms can simply not function. Paying for services would certainly discourage a number of users from utilising the platform. A reduced user base with a possibly diminished interest in the platform may cause the enterprises running such platforms to lose their market domination built upon the network effects.[49] For these reasons, platforms would rather keep their users' attention and time than collect their money. The costs arising from the data collection tax are likely to predominantly be *passed-through onto the customers of the other side(s) of the platform.* From a user perspective, digital services provided in the context of the digital barter are likely to 'remain free' (and this time they would actually become free of charge) or to be sold at prices that are significantly lower than their actual value after the implementation of the data collection tax. Even under this latter scenario, implementing a data collection tax would reveal to be much more beneficial for the overall population.[50]

The second most relevant issue giving rise to a debate would certainly be *discrimination of certain highly digitalised businesses.* It is clear that the data collection tax is, by design, completely neutral. This neutrality is conceived to be not only *de lege* but also *de facto*. In practice, however, the greatest majority of the tax will inevitably be paid by a small number of US companies, referred to by different names such as 'the four' and 'GAFA'. This end result is inherent to the current imbalance of the data market dominated by an extremely limited number of companies. Since it is not possible to fix such an imbalance by means of tax policy or to implement a tax that would unfairly burden other data aggregators to create a certain artificial image of 'fairness' that would further intensify the market

[47] On this point, see Chap. 10.

[48] See Sect. 3.3.1 on the expectation of free digital services.

[49] See Sect. 3.3.3 for more information on the market domination and the 'vicious cycle'.

[50] See the following Sect. 6.7 as well as Chap. 7.

dominance of a handful of companies and perpetuate 'the vicious cycle', it cannot be possibly argued that the data collection tax amounts to an indirect discrimination.

This point can be illustrated by Facebook (Meta) and Google, which seem to be the leaders in data collection. It is likely that a majority of all valuable online, and perhaps even all digital data, are collected by these two enterprises.[51] In 2015, Google's share of Internet searches was 64% and Facebook's 1.5 billion users spent, on average, 50 minutes per day on either the site or the application.[52] The huge fraction of the market controlled by these tech giants entails that they would bear most of the burden of the increased prices for the currently 'free' data,[53] whether due to a data collection tax or not. Such a natural result cannot reasonably be considered as amounting to a discrimination of US companies.

6.7 Allocation of the Funds

Governments would necessarily invest a non-negligible amount to be able to put in place and ensure the effective administration of the data collection tax. Such costs are likely to be much higher during the initial phase of implementation. For under-developed and developing countries, the investment that has to be made will most probably be even more substantial when compared to developed countries with well-trained tax officials, more up-to-date infrastructure and more qualified researchers experienced in the field of data. A part of the proceeds of the tax should, therefore, be allocated to its implementation and administration. This part should, under normal circumstances, diminish with time.

By taking the exponential value of data into consideration as much as possible, the data collection tax should normally generate a non-negligible amount, which would be, in any case, much more consequential than the overall amount that can ever be collected by the users residing in a country in the context of a full-fledged data labour market. In other words, the data collection tax could contribute to the social welfare of a country in a substantially more meaningful manner than an efficient data labour market.

In their work on data dividend, Feygin et al. enumerate four different manners in which the funds may be allocated: state spending on public services, baby bonds or universal saving accounts, *per capita* payments to all users (universal cash payments) and meritocratic payments for data contributions.[54] The contributors recommend the first two options. The authors of this book predominantly agree with this approach, with some, rather minor, discrepancies and additional comments, as detailed below.

---

[51] Posner and Weyl (2019), p. 78.

[52] Posner and Weyl (2019), p. 78.

[53] Posner and Weyl (2019), p. 78.

[54] Feygin et al. (2021), p. 16.

Meritocratic payments for data contributions would necessitate a detailed and thorough examination of data shared by all users. Such an examination will provoke further costs for governments, which may be higher than the additional amounts accruing to a limited number of users. The overall social welfare may, therefore, be reduced. Moreover, scrutinising users' activities to pinpoint 'the users who contribute the most' to data generation may provoke privacy problems.

Per capita payments to all users (universal cash payments) constitutes a viable option. Investing in public services, however, would reveal to be more beneficial for the overall population. Allocating a part of the funds to education and creating an unemployment fund may remedy to a certain extent to the current non-negligible problem of 'disappearing jobs'.[55] Making rather modest *per capita* payments to users may not form a meaningful contribution. Investing a considerable amount to education to ensure that the population gains new skills and to the unemployment fund to sustain individuals losing their jobs until they can re-join the workforce, on the other hand, can.

Baby bonds, which can be described as guaranteed savings accounts at birth, for all residents[56] can also provide an important support to the overall population. Depending on the actual needs of the population based on the relevant economic and social factors, the funds amassed via the data collection tax can be allocated to public services and/or to baby bonds.

Last but not least, a part of the funds should be allocated to cover the costs incurred due to data pollution.[57] Such an investment should also extend to conducting research in this field and implementing measures that would alleviate the negative effects of the phenomenon. Conducting a public campaign to inform users about data generation and collection processes, data breaches and data misuses as well as potential risk factors constitutes an important component of the fight against data pollution.

References

Aslam A, Shah A (2020) Tec(h)tonic shifts: taxing the "digital economy". IMF Staff Discussion Note, IMF

Avi-Yonah RS, Kim YRC, Sam K (2022) A new framework for digital taxation. Harv Int Law J 63(2):279–341

Collin P, Colin N (2013) Task Force on Taxation of the Digital Economy, Report to the Minister for the Economy and Finance, the Minister for Industrial Recovery, the Minister Delegate for the Budget and the Minister Delegate for Small and Medium-Sized Enterprises, Innovation and the Digital Economy

Cui W (2019) The digital services tax: a conceptual defense. Tax Law Rev 73(1):69–11. (electronic version)

[55] On this point, see Chap. 7.

[56] Feygin et al. (2021), p. 17.

[57] For the concept of data pollution, see Sect. 5.2.1.

Feygin Y, Hecht B, Prewitt M et al (2021) A data dividend that works: steps toward building an equitable data economy. Berggruen Institute. https://www.berggruen.org/themencode-pdf-viewer-sc/?tnc_pvfw=ZmlsZT1odHRwczovL3d3dy5iZXJnZ3J1ZW4ub3JnL3dwLWNvbnRlbnQvdXBsb2Fkcy8yMDIxLzA1L0RhdGEtRG12aWRlbmRfZmluYWwucGRmJnNldHRpbmdzPTExMTExMTExMTExMTExMTExJmxhbmc9ZW4tVVM=#page=&zoom=auto&pagemode=. Accessed 7 July 2023

Goulder R (2020) Should data extraction be taxed as a natural resource? Tax Notes Int 27 July 2020: 559–561

IMF (2019) Corporate Taxation in the Global Economy. IMF Policy Paper No. 2019/007, IMF

Marian O (2022) Taxing data. Brigham Young Univ Law Rev 47(2):511–576

Plattner RD (2022) The virtues of a simple excise tax on personal consumer data. Tax Notes Int 108(11):1381–1389

Posner EA, Weyl EG (2019) Radical markets: uprooting capitalism and democracy for a just society, 1st edn. Princeton University Press

Chapter 7
Towards a 'Robo-Data Tax'? (Step 2)

Abstract This chapter aims to examine the ever-increasing technological unemployment phenomenon, with an emphasis on microworkers. The replacement of 'jobs' with 'tasks' and productivity slowdown are briefly described to determine the consequences of the technological unemployment for the society as a whole. Are then succinctly examined, a potential application of robot taxes to highly digitalised enterprises, along with the possibility of implementing a new 'robo-data tax' to alleviate the negative effects of such unemployment.

Microwork, which forms *step two* under the methodology adopted in this book, constitutes one of excessive automation's side effects on the labour market. Thereupon, *the main issue arising in the second step* from a tax policy perspective is the ever-increasing risk of unemployment and the necessity for governments to support their citizens in different manners until the displacement effect can be counterbalanced by other factors.[1] This issue does not have a direct link with taxation of data as such. Nonetheless, microworkers are one of the main actors of the 'data business' and their working conditions[2] demonstrate in a remarkably straightforward manner one of the main problems linked to jobs disappearing due to technological developments. In addition, highly digitalised businesses active in the data market also contribute to the arising unemployment and it should be examined whether it would be suitable to adopt a tax measure specific to them. For these reasons, is being made in this chapter, a brief explanation with regard to unemployment resulting from the fourth industrial revolution along with potential tax policy solutions.

[1] As already mentioned in Sect. 3.5.2.

[2] As elucidated in Sect. 3.3.4.

X. Oberson, A. E. Yazıcıoğlu, *Taxation of Big Data*,
https://doi.org/10.1007/978-3-031-43372-6_7

7.1 Introduction

The starting point of the ongoing discussion can be summarised, in a nutshell, as follows: new technologies are putting people out of work.[3] This phenomenon is being referred to as 'technological unemployment', a term that was introduced by John Maynard Keynes in 1930.[4] The fourth industrial revolution seems to have added an additional layer to the 'traditional' technological unemployment concept, it "[. . .] is not taking certain jobs (artisans, personal assistants who use paper and typewriters) and replacing them with other jobs (assembly-line workers, personal assistants conversant with computers)[;][i]nstead, it is poised to bring about a *wide-scale decimation of jobs* — mostly lower-paying jobs, but some higher-paying ones, too".[5]

A great portion of the current debate seems to evolve around a rather 'simplified' view of the unfolding situation. Such simplified analysis is primarily based on two fallacies 'the lump of labour fallacy' and 'the humanistic fallacy'. The former refers to the conviction that there is a fixed quantity of work with wages above subsistence level; a given 'lump' of labour that is to be divided up and parcelled out either to individuals or to machines.[6] The latter designates "[t]he tendency to assume that the substitution of human work by machines will leave (most) humans with better jobs, than those humans have now".[7] In other words, the former is the pessimistic view and the latter is the optimistic view. As it is developed in this chapter, neither of these two assumptions is accurate. That being said, the humanistic fallacy seems to be outright inaccurate, while the lump of labour fallacy can be corrected with 'minor touches': the quantity of work with wages above subsistence level is being reduced by the automation; excessive investment in automation dilutes the probability of the creation of new reasonably-paid employment; and the current 'jobs' are excessively being replaced by 'tasks', a part of which may be more efficiently[8] carried out by machines.

The purpose of this chapter is neither to argue against technological development nor to imply that cost-saving technologies that could be beneficial to all consumers should not be implemented for the sake of slowing down the automation and saving certain categories of jobs.[9] The discussion is not, and has never been, about saving

[3] See Susskind and Susskind (2017), p. 284.

[4] See, Keynes (1963), pp. 358–373; Susskind and Susskind (2017), p. 284.

[5] Emphasis added by the authors. Lee (2017).

[6] Susskind and Susskind (2017), p. 289.

[7] Loi (2015), p. 202.

[8] In the efficiency analysis, both the volume produced and the costs should be taken into consideration. An example mentioned by Susskind and Susskind illustrates the issue in a very clear manner. A piece of medical equipment ten times more productive than an individual cannot be considered as efficient for doctors to utilise, if the costs incurred by its use are hundred times more expensive when compared to humans [Susskind and Susskind (2017), p. 292].

[9] Moser (2021), p. 494.

jobs at all costs or 'privileging employees over consumers'.[10] One has to take a wider approach to grasp the core issue, which is summarised by Meisel as follows "Automation can reduce personnel costs for companies, but can destroy jobs at too fast a pace for the economy to absorb. Increased productivity can reduce the price of goods, but someone still has to buy those goods. If unemployment remains stubbornly high or even grows, at some point the loss of jobs will overcome the advantage of lower prices, and a long-term recession could ensue".[11]

The aim should be to focus on the current ecosystem as a whole, identify the actual problems and suggest viable policy options. To this end, first the technological unemployment is succinctly detailed and then introduction of a 'robo-data tax' is discussed.

7.2 Technological Unemployment

For the *reinstatement effect* to counteract the *displacement effect* (which directly translates into the technological unemployment) in an efficient manner, the newly created jobs (or alternatively tasks) should not result in dire working conditions and lower wages when compared to the 'old' jobs.[12] Technological unemployment should, thus, be examined by a combination of two interrelated factors: (i) the available *number of jobs* and (ii) the *quality* of the available jobs and, interdependently, the amounts of wages. Irrespective of the arguments put forward by 'the pessimists' and 'the optimists', a number of circumstances currently unfolding are far from being palatable. Such circumstances seem to indicate that both the available number of jobs and the quality of the available jobs may follow a downward trend in the medium to long term.

To be able to reflect the different features of the issue, technological unemployment is examined in four steps. First, the disappearing jobs phenomenon is briefly examined. Second, the productivity slowdown is concisely described. Third, the newly created 'jobs' are compared to the 'old jobs' they are aiming to replace. Finally, the main consequences of the technological unemployment for the society are established.

7.2.1 Disappearing Jobs

The issue of disappearing jobs is already examined in Sect. 3.4.3. This section further develops the concept, without repeating the same information to the extent

[10] *Contra* Moser (2021), p. 493.

[11] Meisel (2013), p. 219.

[12] See also Sect. 3.4.3.

possible, and makes a number of observations that are pertinent for the purposes of the analysis conducted below.

Automation tends to take over 'tasks' and not 'entire jobs'. Consequently, for a job to 'disappear', the full bundle of tasks composing that job should be lost and not replenished with new ones.[13] According to the recurring narrative, routine tasks are more apt to be replaced by automation when compared to non-routine tasks. As a matter of fact, since routine tasks require methodical repetition of an unwavering procedure, they can be exhaustively specified with programmed instructions and performed by machines.[14] By contrast, non-routine tasks cannot be specified in computer code and executed by machines,[15] due to the level of expert thinking, or complex communication of physical tasks they involve.[16] Non-routine tasks tend to contain pattern recognition and fine sensorimotor coordination, which cannot be easily replicated by computer codes (yet).[17] Despite not being apparent at first glance, the terms 'routine' and 'non-routine' used in the context of tasks, do not necessarily correspond to the general understanding of what 'routine' is. For instance, chess, which demonstrates a high level of human skill,[18] constitutes a routine task;[19] whereas navigating a car through city traffic, which tends to be considered as a low-skill and almost 'automated' human activity, forms a non-routine task.[20]

In the light of this general overview, automation would, *prima facie*, only replace routine tasks and humans would concentrate on non-routine tasks of their jobs. Such a collaboration can only give rise to a 'win-win scenario', where overall costs are reduced and employees have more stimulating jobs, free from any repetitive and uninteresting tasks. As one may suspect, this overly optimistic scenario seems to dissipate when the issue is examined in further detail. In fact, many apparently non-routine tasks can be routinised, and many genuinely non-routinisable tasks can still be undertaken by machines.[21] This renders the occurrence of a significant shift of tasks away from humans towards machines highly probable.[22] In addition, as machines become increasingly capable, it is far from being certain that individuals may secure their place indefinitely in the human-machine joint ventures.[23]

Automation instruments are increasingly performing functions that were previously fulfilled by humans, reducing, thereby, the need for certain types of workers.

[13] Susskind and Susskind (2017), p. 287.

[14] Autor et al. (2003), p. 1283.

[15] Autor et al. (2003), p. 1283.

[16] Loi (2015), p. 206.

[17] Loi (2015), p. 206.

[18] Chess is also a recognised sport.

[19] Brynjolfsson and McAfee (2014), p. 190.

[20] Autor et al. (2003), p. 1283.

[21] Susskind and Susskind (2017), p. 291.

[22] Susskind and Susskind (2017), p. 291.

[23] Susskind and Susskind (2017), p. 293.

In a simultaneous manner, however, the demand for more sophisticated human functions, like data analysis, is rising.[24] Also, a number of jobs involving a certain degree of sensorimotor work, in some occasions along with the skills of ideation, large-frame pattern recognition and complex communication, such as cooking, gardening, repairing and providing health aid, are 'machine-replacement resistant', at least for a while.[25] Furthermore, individuals are likely to retain the advantage over machines in tasks involving making or supporting important moral decisions,[26] such as conceding custody in a divorce action, and positively discriminating in favour of a minority applicant to a university.[27]

That being said, there is a mismatch between numbers of jobs that are disappearing and numbers of jobs that are being created. Individuals demonstrating high skills in new areas may be needed, but not in numbers as great as individuals losing their jobs.[28] Moreover, as a natural result of vanishing middle-class jobs, people who were previously engaged in middle-skill knowledge work expand their job search to low-skill jobs.[29] This, in turn, places a stress in low-skill jobs that are not otherwise in a 'race with machines' and causes a downward trend both on available employment opportunities and wages.[30] In short, jobs are disappearing and a significant shift may occur in the labour market in the medium term, mostly affecting low and middle-skilled jobs.

In the long term, even the replacement of *all* human tasks by automation can become possible. As stated by Brynjolfsson and McAfee, "[...] watching one bastion of human uniqueness after another fall before the inexorable onslaught of innovation, it's becoming harder to have confidence that any given task will be indefinitely resistant to automation".[31] The contributors further add that "[...] it's very easy to underestimate the power of digital, exponential, and combinatorial innovation. So never say never".[32]

7.2.2 *Productivity Slowdown*

The displacement effect occurring in the labour market could be counterbalanced by different factors, among which productivity. The rapid advances in technology,

[24] Loi (2015), p. 204.

[25] Brynjolfsson and McAfee (2014), p. 202.

[26] Susskind and Susskind (2017), p. 291.

[27] Susskind and Susskind (2017), p. 281.

[28] This phenomenon of disappearing jobs is examined in detail and with concrete examples in Sect. 3.4.3.

[29] Brynjolfsson and McAfee (2014), p. 202.

[30] Brynjolfsson and McAfee (2014), p. 202.

[31] Brynjolfsson and McAfee (2014), p. 203.

[32] Brynjolfsson and McAfee (2014), p. 204.

however, did not translate into a high level of productivity. Quite the contrary, productivity has slowed down in a considerable manner since 2004.[33] Initially, it was argued that a real slowdown did not exist, but the rise of intangibles simply meant that old metrics were mismeasuring productivity and investments by firms into new technology.[34] This once popular theory has been proven wrong.[35]

The productivity slowdown can be explained by a cumulation of different reasons. Undisputedly, one of the factors is the almost singular focus in the corporate sector and research community on AI, applications of deep learning and other 'big data methods' to automate various tasks.[36] Too much attention tends to be devoted to automation at the expense of other technological breakthroughs.[37] This *excessive automation tendency* acts as a drag on productivity growth.[38] Another relevant cause is the concentration of new technologies at the hands of siren servers, which use their advantages to strengthen their dominance in the market and to prevent technology from diffusing to competitors.[39] Such a level of concentration accounts for a part of the decrease in 'wage share'—i.e., share of profits accruing to workers—.[40] A further reason is the lack of skills of the workforce. If certain skills are complementary to new technologies, their absence will imply that the productivity of the technologies concerned will remain lower than it could have been.[41] This point seems to have been demonstrated by the productivity growth of the last two decades. The existent growth has predominantly come from workers' investments in their skills.[42]

Without a strong productivity effect, it would not be feasible to create enough jobs to tackle the technological unemployment. In the long term, new developments could occur and unemployment may not even remotely be a problem. That being said, as demonstrated by other similar examples in history, a burdensome transition period would be necessary. The main question is how far should excessive automation be taken with the hope of future progress. In other words, what threshold of intergenerational injustice, privileging future generations over the generations that will have to go through the painful transition period, should be deemed appropriate?[43]

---

[33] Feygin et al. (2021), p. 10.
[34] Feygin et al. (2021), p. 10.
[35] Feygin et al. (2021), p. 10.
[36] Acemoglu and Restrepo (2018), p. 15.
[37] Acemoglu and Restrepo (2018), p. 15.
[38] Acemoglu and Restrepo (2018), p. 3.
[39] Feygin et al. (2021), p. 10.
[40] Feygin et al. (2021), p. 10.
[41] Acemoglu and Restrepo (2018), p. 13.
[42] Feygin et al. (2021), p. 10.
[43] See also, Loi (2015), p. 202.

7.2.3 *Jobs Are Being 'Replaced', for Better or for Worse?*

> The lightening of the labour, even, becomes a sort of torture since the machine does not free the labourer form work, but deprives the work of all interest.[44]

New technologies have often displaced workers but new opportunities have always arisen to meet new demands.[45] The same is likely to hold true for the current technological unemployment. As machines get more powerful and algorithms become able to solve a greater number of problems, industries will identify new issues to tackle.[46] In other words, the frontier between what machines can and cannot achieve is ever-moving.[47] This phenomenon can be referred to as the '*paradox of automation's last mile*'; every progress made by machines gives rise to additional automation opportunities.[48] The constant repetition of this process results in the expansion of automation through the perpetual creation and destruction of labour markets for new types of human labour.[49] Human labour is, thus, shifting its shape to fit the new environment and not being entirely replaced. What needs to be established is whether the new shape would be 'better' or 'worse' than the older. Would most humans end up with better jobs than the ones they have now, as argued by the humanistic fallacy?[50]

To answer to that question a comparison needs to be made between jobs most humans undertake now (with the existent technology) and jobs most humans will undertake in the future (with the technology that may exist in the future).[51] Would most future jobs indeed be less boring, repetitive, exhausting, directly controlled by others and dangerous, than most jobs that exist now?[52] And, perhaps even more importantly, would they give rise to higher wages or at least wages comparable to the current ones?

These questions, it would seem, have to be answered by the negative. As underlined by Susskind and Susskind, new services, and conceivably even new industries, may be created as a result of automation, just like it was the case in the past.[53] There exists, however, one fundamental difference: new job opportunities were created in the past not due to the fact that new services could be provided, but due to the fact that provision of the services concerned required new tasks to be performed in which *individuals*, and not machines, *had the advantage*.[54] While it is

---

[44] Marx (1887), p. 286.

[45] Susskind and Susskind (2017), p. 293.

[46] Gray and Suri (2019), p. 184.

[47] Gray and Suri (2019), p. 184.

[48] Gray and Suri (2019), p. 184.

[49] Gray and Suri (2019), p. 184.

[50] Loi (2015), p. 202.

[51] Loi (2015), p. 205.

[52] Loi (2015), p. 205.

[53] Susskind and Susskind (2017), p. 294.

clear that new services will be demanded in the future, it would not be correct to assume that individuals would systematically have the advantage over the machines in the exercise of such services.[55] On this point, Gray and Suri project that at the current rate of growth combined with current trends in the growth of contract staffing and temporary agency services, 60% of today's global employment is likely to be converted into some form of microwork by 2055.[56] Should that outcome effectively materialise, humans would indeed be left with boring, repetitive, exhausting, directly controlled by others jobs, with wages considerably below the subsistence level.[57] Stating that the new tasks created by the microtask industry could replace the current jobs and form a viable post-automation employment option would be, to say the least, an overreaching argument.[58]

The expansion of microwork would leave the workers with 'tasks' that need to be fulfilled under dire working conditions, combined with little compensation and, thereby, lower purchasing power. On top of such predicaments, repetitive and exhausting jobs where humans are stripped out of their personal identities, would clearly reduce a number of human abilities and may cause health problems in the long-term.[59] Also, as mentioned above,[60] individuals currently exercising middle-skill jobs are likely to turn to low-skill jobs as a result of the technological unemployment. It can, therefore, be stated that labourers face being 'disenhanced' as a side effect of automation, rather than being 'enhanced'.[61]

7.2.4 Consequences of the Technological Unemployment for the Society as a Whole

Beyond any doubt, technological unemployment is not a phenomenon affecting merely an isolated group of individuals 'losing their jobs'. It rather produces a significant impact on the society as a whole. With their citizens facing unemployment, and possibly health problems linked to stress or 'new' dire working conditions, governments will have to intervene to reduce the impact of automation in the labour market and to restore, as much as possible, a certain level of balance. Reforming the educational system, putting in place specialised skill developing programs for the unemployed, investing in research to enable technological

[54] Susskind and Susskind (2017), p. 294.

[55] Or vice versa. Susskind and Susskind (2017), p. 294.

[56] Gray and Suri (2019), p. 179.

[57] For more details on working conditions of the microworkers, see Sect. 3.3.4.

[58] See Sect. 3.4.3.

[59] Repetitive, exhausting jobs that are predominantly controlled by others may produce negative effects on health. On this point, see Loi (2015), p. 205.

[60] In Sect. 7.2.1.

[61] For the concept of 'human disenhancement', see Loi (2015), pp. 203–204.

breakthroughs to raise productivity and provision of a Universal Basic Income to all citizens under challenging economic circumstances to alleviate the side effects of automation all constitute viable potential solutions.

Determining the most convenient types of government intervention to tackle technological unemployment is beyond the scope of this book. Regardless of the method(s) chosen by a given government, one thing is certain: the funds that will need to be allocated to alleviate the effects of technological unemployment until the transition period is completed, are substantial. To ask the obvious, how can governments finance such substantial sums in a sustained manner over a non-negligible time frame?

The main problem is that technological unemployment not only necessitates an additional investment from the government to support the labourers and the economy, but also reduces the financial resources of the government in a substantial manner. Such reduction mainly occurs due to two reasons. First, the wide-scale decimation of jobs necessarily translates into the contraction of revenue accruing from income taxes, and other similar levies, collected on the basis of salaries. Second, consumption also diminishes as a result of disappearing income of workers, which translates into less tax revenues deriving from consumption taxes and, even more importantly, a shrinking economy. Availability of cheaper products and services may not mean much to a population, a considerable part of which is below the subsistence level.

The end result is governments need extra funds to be able to cope with a large-scale technological unemployment. Tax policy could certainly offer a certain degree of support. For instance, a part of the funds generated by means of data collection tax[62] could be employed to subsidise the measures put in place to attenuate the impact of the unemployment. Another possibility open to discussion is the implementation of a 'robo-data tax'.

7.3 Robo-Data Tax

Various tax policy suggestions to alleviate the effects of the technological unemployment have been made in the last decade, the most popular of which is the implementation of a 'robot tax'.[63] The term 'robot' is to be interpreted in a broad and form-neutral manner, with an emphasis on the ability to process, plan and act on its own, comprising, therefore, AI.[64] The proposal to tax robots suggests attributing (imputed) income either to the use of robots by enterprises or to robots as such.[65] Under the first and general proposal, the robot tax concentrates on the use of robots

[62] See Chap. 6.

[63] See, for example, Oberson (2019).

[64] Oberson (2019), pp. 15–16.

[65] Oberson (2019), pp. 112–142.

and is levied from *enterprises* utilising such robots in order to compensate for the disappearing human labour. The second, and more provocative, proposal goes one step further and suggests that the legislator could even contemplate granting a tax personality to *robots or AI, as such*, and, as a consequence, levy a specific tax from 'robots' as the subjects[66] of the tax.[67]

Should a tax on the *use of robots* be effectively implemented, highly digitalised enterprises active in the data market are most likely to be submitted to such a tax in any case, i.e., without the need to incorporate any specific provisions to that end. In fact, as described throughout this book,[68] big data utilised by such enterprises are collected, processed and analysed in a largely autonomised manner via algorithms and AI. In the scenario where a robot tax is levied by means of granting a *tax personality to AI* and collecting a specific tax from this new category of taxpayer, for highly digitalised enterprises active in the data market to be submitted to tax, algorithms used to collect, process and analyse data should be granted a tax personality. This does not give rise, however, to a special treatment of the enterprises active in the data market, since all highly digitalised enterprises would be subject to a similar practice, as dictated by the rationale of such a tax. Sector-specific rules and regulations may, nonetheless, reveal to be necessary to be able to implement such a tax, which aims to introduce a new 'fiction' into the legal realm along with all the complexities it may bring.

Whether the implementation of *a sector specific robo-data tax* would reveal to be a viable policy option may provoke a fascinating debate. To be precise, so far, the rationale of taxing AI or robots has not been motivated by taxing data in the most efficient manner possible. A robot tax, in general, focuses on the profits arising, either (i) from the use of robots instead of human workers; or (ii) the activities of robots, to the extent they would be granted some kind of tax capacity. Under the Pigouvian perspective, a robot tax may be justified by the accentuated need of compensation for the negative externalities provoked by the unemployment caused by automation. The robo-data tax discussed here as a potential policy option, on the other hand, consists of a tax levied on the collection of data by algorithms and AI. It is, therefore, sector specific. Such a tax would focus more on the *input side*, namely on data considered as information generating value for future profits of enterprises (i.e., raw material); in contrast to a robot tax, which aims to apprehend the profits or income at the *output side* for transactions undertaken by 'robots'.

The transformation that would ultimately give rise to a wide-scale technological unemployment is the same transformation that creates siren servers and that results in substantial profits for the enterprises collecting tremendous amounts of data, analysing the data concerned and providing their services on that basis through the use of their AI.[69] The majority of such enterprises are residing either in the US or in

[66] In other words, the taxpayers.

[67] Oberson (2017), p. 254.

[68] Especially in Chaps. 2 and 3.

[69] See also, Lee (2017).

China. In view of the current circumstances and trends, it cannot be excluded that, in a not very far future, US businesses will dominate developed as well as some developing AI markets, while Chinese companies will dominate most developing AI markets.[70] If such a domination indeed occurs, it will lead to serious concerns, which are summarised in the following manner by Singh and Vipra "developing countries [would be left] fully digitally dependent on US or Chinese corporations in almost all sectors, taking their citizens down collectively".[71] Lee makes a similar observation, by indicating that, unless countries are able to tax the 'ultra-profitable' AI companies to subsidise their workers, they will have only one option to prevent plunging their citizens into poverty: negotiating with whichever country supplies most of their AI software—i.e., China or the US—to "[. . .] essentially become that country's economic dependent, taking in welfare subsidies in exchange for letting the 'parent' nation's A.I. companies continue to profit from the dependent country's users".[72]

When such scenarios become foreseeably relevant in the light of robust scientific research and thorough analysis, the implementation of a robo-data tax may be discussed to prevent such an undesirable outcome. Or, the 'welfare subsidies' may take the form of a 'robo-data tax' for a more politically correct implementation. In the current state of affairs, however, this discussion seems exceedingly premature from a tax law perspective, especially when the myriad of legal and political problems that will be caused by it are taken into consideration. Also, in the authors' opinion, an efficient implementation of a data collection tax[73] combined with a proper international allocation of taxing rights (in relation to income taxes)[74] would contribute to solve the problem, at least partially, without radically forcing the current boundaries of the tax systems.

In any eventuality, levying a robot tax or a robo-data tax would inevitably distort the market to a certain extent and should be designed in a cautious manner, taking into account not only the technological unemployment but also many other relevant factors, such as the industrial development level of the country concerned (especially when compared to other countries), immigration inflows,[75] potential effects on consumers, political relations as well as international collaboration.

[70] See also, Lee (2017).

[71] Singh and Vipra (2019), p. 53.

[72] See also, Lee (2017).

[73] See Chap. 6.

[74] See Chap. 8.

[75] Chand et al. (2020).

References

Acemoglu D, Restrepo P (2018) Artificial intelligence, automation and work. NBER Working Paper Series No. 24196, National Bureau of Economic Research

Autor DH, Levy F, Murnane RJ (2003) The skill content of recent technological change: an empirical exploration. Q J Econ 118(4):1279–1333

Brynjolfsson E, McAfee A (2014) The second machine age, 1st edn. W. W. Norton & Company

Chand V, Kostić S, Reis A (2020) Taxing artificial intelligence and robots: critical assessment of potential policy solutions and recommendation for alternative approaches, sovereign measure: education taxes/global measure: global education tax or planetary tax. World Tax J IBFD 12(4): 1–36

Feygin Y, Hecht B, Prewitt M et al. (2021) A data dividend that works: steps toward building an equitable data economy. Berggruen Institute. https://www.berggruen.org/themencode-pdf-viewer-sc/?tnc_pvfw=ZmlsZT1odHRwczovL3d3dy5iZXJnZ3J1ZW4ub3JnL3dwLWNvbnRlbnQvdXBsb2Fkcy8yMDIxLzA1L0RhdGEtRGl2aWRlbmRfZmluYWwucGRmJnNldHRpbmdzPTExMTExMTExMTExMTExMTExJmxhbmc9ZW4tVVM=#page=&zoom=auto&pagemode=. Accessed 7 July 2023

Gray ML, Suri S (2019) Ghost work: how to stop Silicon Valley from building a new global underclass, 1st edn. Harper Business

Keynes JM (1963) Essays in persuasion, economic possibilities for our grandchildren, 1st edn. W. W. Norton & Company

Lee K (2017) The Real Threat of Artificial Intelligence, The New York Times. https://www.nytimes.com/2017/06/24/opinion/sunday/artificial-intelligence-economic-inequality.html. Accessed 7 July 2023

Loi M (2015) Technological unemployment and human disenhancement. Ethics Inf Technol 17: 201–210

Marx K (1887) Capital volume one: the process of production of capital. Progress Publishers, Moscow. https://www.marxists.org/archive/marx/works/download/pdf/Capital-Volume-I.pdf. Accessed 7 July 2023

Meisel W (2013) The software society: cultural and economic impact. Trafford Publishing

Moser E (2021) Against robot taxes: scrutinizing the moral reasons for the preservation of work. AI Ethics 1:491–499

Oberson X (2017) Taxing robots? From the emergence of an electronic ability to pay to a tax on robots or the use of robots. World Tax J IBFD 9(2):247–261

Oberson X (2019) Taxing robots: helping the economy to adapt to the use of artificial intelligence, 1st edn. Edward Elgar Publishing

Singh PJ, Vipra J (2019) Economic rights over data: a framework for community data ownership. Development 62:53–57

Susskind R, Susskind D (2017) The future of the professions: how technology will transform the work of human experts, Updated edn. Oxford University Press

Chapter 8
Permanent Establishment: The Fairest Nexus of All? (Step 3)

Abstract In this chapter, the authors argue that the current provisions of double tax treaties amount to a ring-fencing of the traditional economy. To remedy to this problem, the authors suggest the creation of a new 'virtual permanent establishment' nexus, combined with the addition of a new category of income, namely 'digital services fees'.

8.1 Introduction

> The current system is a compromise, and the future system will be as well.[1]

With regard to source taxation, the crux of the debate relating to taxation of highly digitalised enterprises concentrates on taxation of multi-sided platforms. In fact, while specificities of digital assets and services may necessitate further reflection for tax purposes, the business model followed by multi-sided platforms requires a fundamental shift in the current international tax system.

Multi-sided platforms tend to employ data collected from a user base situated in one country to render services to their customers on the other end(s) of the platform in relation to the users of the country concerned. For instance, a multi-sided platform operating a search engine is going to rely on the information that it has on the users located in Country A to render advertisement services to companies (which may or may not be residents of Country A) wishing to advertise to their potential clients in Country A. To further concretise the example, one may think of the advertisements of French and German car manufacturers provided via Google's search engine to users residing in France.

As is common knowledge, a certain threshold needs to be met for source-based taxation (i.e., taxation of non-residents on income derived from a country) to occur. Such threshold may derive from the domestic law of the source country and/or from the international tax law, which comprises DTTs and other international instruments ratified by the country concerned. Any domestic law threshold that is larger than the

[1] Schön (2018), p. 280.

X. Oberson, A. E. Yazıcıoğlu, *Taxation of Big Data*,
https://doi.org/10.1007/978-3-031-43372-6_8

threshold designated by the international tax law will be subject to the limitations established by the latter. As illustrated by means of a court decision in the report prepared by the French Task Force on Taxation of the Digital Economy,[2] with regard to the business model used by multi-sided platforms, the contours of the threshold employed in DTTs are likely to restrict the national thresholds, which tend to cast a wider net. As per the decision rendered by the Council of State[3] in 1968, the profits of a company headquartered in the Principality of Monaco that took orders from French clients for radio advertising messages that were broadcast to French listeners were taxable in France, due to the fact that the company conducted its 'complete business cycle' in France and should, thereby, be considered as a 'company operating in France' under the General Tax Code.[4] The report underlines that the same radio station broadcasting to France from Ireland and signing contracts with its advertisers from Ireland would not be required to pay corporate income tax in France because of the 'systematic inclusion of clauses' in DTTs that grant the right to tax to the residence state of the company concerned.[5]

To determine the international tax law threshold that must be met to allow the source taxation of a multi-sided platform providing digitalised services, first the income obtained by such platforms should be categorised under the provisions of the OECD MC and the UN MC. In the current state of the Model Conventions, the broadest and the primary category that would typically apply is business profits (article 7), which necessitates as a minimum threshold the existence of a PE for a source taxation. Depending on the circumstances of a given case, other categorisations that may reveal to be most appropriate are royalties under both Model Conventions; as well as technical services fees and income from automated digital services under the UN MC. To the extent where royalties, technical services fees and income from automated digital services are effectively connected with a PE situated in the source country, article 7 remains applicable. In the eventuality where the PE threshold is not reached but business profits in the form of royalties, technical services fees or income from automated digital services are being derived; article 12 of the OECD MC and the UN MC, article 12A of the UN MC or article 12B of the UN MC becomes applicable. While the UN MC grants a right to tax to the source country in all three cases, the OECD MC grants the exclusive right to tax the royalties to the residence state.

As examined in this chapter, the existing rules of the DTTs do ordinarily not allow source countries to submit to tax multi-sided platforms providing digital services. This systematic lack of taxation in source countries has been the starting point of the debate on the digitalisation of the economy. While it is not feasible to provide a comprehensive overview of all the discussions on the issue along with all the suggestions that have been made so far, the key problems resulting in the absence

[2]Collin and Colin (2013).

[3]The supreme tax jurisdiction of France.

[4]Collin and Colin (2013), p. 25.

[5]Collin and Colin (2013), p. 25.

of source taxation and the most appropriate solution to these problems according to the authors are briefly explained in the following sections. Prior to begin the analysis, a number of observations need to be made and some basic concepts need to be reminded.

As expressed by Arnold, the taxation of non-residents' business profits under both domestic laws and international tax treaties consists of six basic stages.[6] The *first stage* is to identify the circumstances under which a country has the right to tax non-residents. While it is debatable whether this initial question of nexus is determined on the basis of a theoretical justification for taxing non-residents or the limits on the practical ability of countries to enforce taxes on non-residents; as stated by Schön, the current debate on taxation of the digital economy seems to concentrate on accommodating the competing tax claims of the countries involved in a manner that is administrable from a practical point of view and acceptable from a revenue point of view, rather than developing a new consensus based on "hard-wired legal considerations or scholarly public finance concepts like efficiency or neutrality".[7] This point is also emphasised by Jiménez, who describes the debate on the taxation of digital economy as a "mixture of economics and legal reasoning together with policy and politics".[8] The *second stage* consists of defining 'business' under domestic laws as well as DTTs and distinguishing between business profits and investment income. The *third stage* involves determining the relevant threshold that needs to be exceeded for non-residents to become taxable in the source country. It is also possible to submit to tax all profits derived from business carried on in the country concerned, without any specific minimum thresholds. The *fourth stage* consists of establishing the rules (commonly referred to as geographic source rules), which are necessary to determine which business profits derived by a non-resident from its business are attributable to and taxable by the source country. The *fifth stage* encompasses the determination of the rules to be applied for the purposes of computing the business profits derived by a non-resident. DTTs tend to rely on domestic laws to provide detailed computational rules, subject only to broad principles of non-discrimination, separate accounting and the arm's length standard. The *sixth and the final stage* comprises the establishment of the rules that regulate the payable tax and its collection. Non-residents are typically subject to rules that differ from those applicable to residents, given the greater difficulty in collecting tax from the former category of taxpayers. These six stages are intimately connected.

This chapter briefly discusses the relevant nexus, the minimum threshold needing to be exceeded for a source taxation as well as problems that may occur during the computation of business profits. It is clear that the solution suggested by the authors, which essentially consists of a combination of proposals that have already been made by different institutions and contributors with slight modifications, is not ready-to-implement. It necessitates a thorough examination of a myriad of issues

[6] See Arnold (2003), p. 477.
[7] Schön (2018), p. 280.
[8] Jiménez (2018), p. 621.

and a much-detailed work. The developments made in this chapter can serve as a contribution to the on-going debate. It is also beyond discussion that the OECD has already attempted to remedy to the issues arising from source taxation of digitalised services by means of the MLI and the Two-Pillar Approach.[9] In the authors' opinion, however, a more efficient and an easier-to-implement solution does exist. This observation is also due to the fact that the tax analysis of this step (*step three*) effectuated by the authors builds on the data collection tax.[10] In other words, when analysing the potential tax implications of digital services rendered by multi-sided platforms, the authors assume that a data collection tax was already levied in *step one*.

The implementation of a data collection tax will produce a major outcome for the purposes of the tax analysis conducted with regard to *step three*: by ensuring that an appropriate fee has been paid in exchange of the raw material necessary to conduct the digitalised activities, it alleviates to a great extent the need to take the collection of raw data into consideration in the value chain and, thereby, reduces the complexity.

The contours of the concept of value creation are quite vague.[11] Within the context of the digital economy, value creation seems to be interpreted as the fact that an enterprise's value is enhanced not only by its supply side but also by the market itself.[12] This interpretation is supported by means of several factors, such as network effects, content created by users and data generated by users. This interpretation, however, also gives rise to a number of challenging discussions, over which no consensus seems to have been reached. One of the main arguments that has frequently been invoked is the assumption that while the collection of raw data may represent a certain significance, the real 'value creation' occurs in the country where the data are being processed and the services are being prepared. This argument, along with other similar assertions, seems to have resulted in a misleading opinion on the plea for a source taxation of the income deriving from digital services. Under this line of reasoning, the countries arguing for the necessity of a source taxation "appear to be moving to define new ways of taxation, whereby taxation should not occur where value is created, but, rather, where the value is consumed, i.e., the market jurisdiction".[13] Accordingly, the reference is frequently made to market country (and not to source country) in the current debate on the digital economy.

In the current scheme of things, users are indeed assuming the role of 'unconscious contributors' to the business value of a highly digitalised company, almost as if they are the 'unconscious employees' in the company's value chain.[14] It is well-established that data form an important contributor to the value creation. Users are

---

[9] As developed in Sect. 5.1.2.

[10] Data collection tax is examined in Chap. 6.

[11] On this point, see, for example, Elliffe (2021), pp. 34–41; and Olbert and Spengel (2017), p. 5.

[12] Hongler and Pistone (2015), p. 3.

[13] Petruzzi and Buriak (2018), p. 4.

[14] Petruzzi and Buriak (2018), p. 14.

generating a significant amount of data while 'consuming' the 'free' digital services provided by highly digitalised enterprises. These data, without which a non-negligible number of services cannot even materialise, are diffused in multi-sided business models, where they are employed for a variety of different purposes. In the face of this pattern, the issue of adequate taxation inevitably arises along with the question of how to attribute value to the generation, storage and use of data.[15]

One of the major challenges in the 'digital economy' debate is thus to understand the manners in which data are monetised and to pinpoint the functions that are involved in that process.[16] Such understanding would serve two purposes: (i) being able to determine whether the remote collection of data should give rise to a taxable nexus on the basis of a thorough analysis of functions performed, assets used and risks assumed in order to reflect the value chain of businesses leveraging data; and (ii) being able to establish the most appropriate TP method(s) to allocate respective profits to the countries where the data are collected, processed and used.[17]

A data collection tax levied during the phases of data generation and collection produces the effect of simplifying the current debate to a great extent. From a purely mechanical standpoint, the generation of data by users and the collection of that data by companies will not have to be taken into consideration while determining the most appropriate way to submit the highly digitalised companies' profits to tax. The most notable effect will occur, however, on the basis of the change of paradigm that will be provoked by the implementation of such a tax. The merit of a data collection tax is not limited to generation of tax revenue. That amount may even reveal to be lower than the proceeds of a corporate income tax put in place on the shoulders of a complex allocation mechanism. The primary achievement of a data collection tax would be to put in motion a shift in the overall perception of the 'data business'. Once the raw data begin to be compensated in monetary terms by the companies acquiring them, the users will not be considered as assuming the role of 'unconscious contributors' or serving as 'unconscious employees'. The highly digitalised companies will not be regarded as entities not paying their 'fair share of taxes' and benefitting 'unjustly' from their consumers. Last but not least, the need to take the data generation and the role of users in such generation into consideration while determining the appropriate corporate income tax treatment will be lifted. The generation and collection of data will become no different than the purchase of another raw material necessary for the production of the goods and services commercialised by the company concerned (with a slight reservation on the non-existence of a free market of data, at least for a certain period of time until such a market could become fully functional). On that basis, a healthier debate can begin and a much simple to understand, easy-to-administer and effective solution can be reached.

---

[15] Olbert and Spengel (2017), p. 5.

[16] Olbert and Spengel (2017), p. 5.

[17] Olbert and Spengel (2017), p. 5.

It is clear that the levy of a data collection tax does not constitute a panacea able to solve all the interpretation problems relating to the digital services rendered by multi-sided platforms. It constitutes just a starting point, from where a more adequate analysis and understanding of the current data business can be effectuated. The argument that source countries aiming to implement measures to enable source taxation of multi-sided platforms providing digital services to customers (i.e., companies situated on the other end(s) of the platform, such as advertisers) are trying to levy a tax in the market country where the value is not created but only consumed does not constitute an appropriate interpretation of the current real-life functioning of the data business. Multi-sided platforms render services that are specific to the user base situated in a given country with the raw material obtained from that country. The services they render should be tailored to the country in which they are used. Services that are custom-made for Country A cannot possibly be employed without any modifications in Country B. In other words, multi-sided platforms' services are tightly linked to the location in which they are used. This link may reveal to be much stronger than the connection that a company active in the traditional economy has with a source country. The same computer, the same furniture and the same good can be sold in any market jurisdiction; on the other hand, a service fabricated on the basis of the information that is locally obtained from a certain country can only be commercialised in relation to that particular country. The link displayed between multi-sided platforms and source countries is also stronger when compared to enterprises active in the digital economy following other business models. For instance, a software or a cloud computing service may be commercialised, with minor modifications, in different countries, unlike a service provided by a multi-sided platform, which is unique to the source country. On the basis of this non-negligible difference, the general opinion that seems to reign in the current debate should not be applied to multi-sided platforms. Countries in which such platforms are rendering their services cannot be assimilated to mere market jurisdictions, where no value is created and only consumption occurs. The contribution provided by these countries is much more pronounced than the contribution provided by a typical source country within a traditional economy framework.

Undoubtedly, this new business model necessitates a shift in the traditional perception of the concept of 'source country'. This is due to the fact that under the 'physical' economy model, services need to be performed in the country where the consumers/customers are residing, whereas under the 'digital' economy model no physical presence is needed in the country where the consumers/customers are residing. This absence of physical presence does not entail that the source country is not contributing to the creation of the services concerned, in other words to the creation of value. For these reasons, the authors of this book opt for using the term of source country and not market country.

Building on the data collection tax proposed in *step one*, the authors are suggesting the implementation of a new type of PE ('virtual PE', which has also been referred to as 'digital PE'), combined with a 'residual' category of 'digital services fees'. As explained in the following sections in details, this proposal does not exactly constitute a novelty, since both concepts have previously been suggested by a number of institutions, including the OECD, and scientific contributors, with some rather minor adjustments made by each institution and scientific contributor. The authors are re-examining these issues in the light of the developments made throughout the previous chapters, to put forward their own point of view on the source taxation conundrum of digital services rendered by multi-sided platforms. With this purpose in mind, this chapter is divided into two sections. First are described the current provisions of the Model Conventions, namely PE, royalties, technical services fees and income from automated digital services, along with the relevant thresholds for source taxation. This examination is kept as brief as possible, due to the existing general knowledge of the provisions and the thresholds concerned. Then, the suggested new nexus is examined in greater details.

8.2 Current Provisions of the Model Conventions

The existing income categories that may potentially be applied to digital services are business profits (article 7 of the OECD MC and the UN MC -which is applied in conjunction with article 5 of the OECD MC and the UN MC regulating the PE-), royalties (article 12 of the OECD MC and the UN MC), technical services fees (article 12A of the UN MC) and income from automated digital services (article 12B of the UN MC). These categories are briefly described one by one to pave the way for the argumentation and observations made in relation to the new nexus suggested by the authors.

8.2.1 Concept of PE

"The PE concept is obviously the most important threshold for source-country taxation of business profits".[18] In accordance with the allocation rule contained in article 7(1) of both the OECD MC and the UN MC, the business profits of an enterprise of one contracting state are only taxable in that state, unless the enterprise concerned carries on business in the other contracting state through a PE situated therein. In the eventuality where a PE is established in the latter contracting state (i.e., the source country), not only does the state concerned acquire the right to tax the business profits attributable to the PE, but its right to tax also takes precedence over the former contracting state's (i.e., the residence country) right to tax such profits.[19]

[18] Arnold (2003), p. 478.

[19] Arnold (2003), p. 478.

The existence of a PE is determined in accordance with article 5 of the OECD MC and the UN MC. The two main types of PE consist of 'fixed place PE'[20] and 'dependent agent PE'.[21] The third, and less frequently encountered type of PE, is 'service PE'.[22] Regardless of their differences, all these three types of PE necessitate the existence of a 'physical presence' in the source country.

The 'physical presence' requirement becomes most apparent in the context of *the fixed place PE*, which relies upon a place of business that is fixed. A place of business may be defined as any premises, facilities or installations used for carrying on the business of the non-resident enterprise, regardless of whether such place of business is exclusively used for that purpose or not[23] and whether the enterprise concerned has a formal legal right to use the place or not.[24] It follows that place of business is a concept that is interpreted broadly, varying from a factory[25] to 'a certain amount of space'[26] that is at the disposal of the non-resident enterprise. To be considered as 'fixed', a place of business needs to be located at a specific geographical nexus[27] and must not be of a purely temporary nature.[28] The requirement of being geographically 'fixed' does not imply that the place of business should be literally fixed to the soil.[29] Underground facilities[30] and movable place of businesses[31] may also constitute 'fixed' places. To constitute a PE, a movable place of business should move around within a geographically coherent area and perform a systematic commercial activity, like a market-place.[32] Regarding the relevant time threshold, although most countries consider a place as fixed only if it is maintained for at least 6 months,[33] a PE may also be deemed to exist in cases where the place of business is kept for a very short period of time because the nature of the business is such that it will only be carried on for that short period of time.[34] This may typically

[20] Article 5(1) of the OECD MC; Article 5(1) of the UN MC.

[21] Article 5(5) of the OECD MC; Article 5(5) of the UN MC.

[22] Article 5(3)(b) of the UN MC. An optional service PE provision is also provided for in the OECD Commentary (2019) ad article 5, §144.

[23] OECD Commentary (2019) ad article 5, § 10.

[24] OECD Commentary (2019) ad article 5, § 11.

[25] As mentioned in (d) of the non-exhaustive list of examples provided by article 5(2) of the OECD MC.

[26] OECD Commentary (2019) ad article 5, § 10.

[27] OECD Commentary (2019) ad article 5, § 21; Sasseville and Skaar (2009), pp. 25–26.

[28] OECD Commentary (2019) ad article 5, § 6.

[29] Sasseville and Skaar (2009), p. 26.

[30] Like pipelines and mines.

[31] Such as a transportable newsstand, a tent or a van equipped for selling products [Sasseville and Skaar (2009), p. 26].

[32] Sasseville and Skaar (2009), p. 27.

[33] OECD Commentary (2019) ad article 5, § 28; Sasseville and Skaar (2009), pp. 29–30.

[34] OECD Commentary (2019) ad article 5, § 28.

be the case for recurrent activities and for activities exclusively carried on in the country concerned.[35]

A nexus for business income based on the concept of fixed place of business has several advantages, namely certainty, enforceability and limitation of fragmentation.[36] The certainty derives from the fact that the existence of a fixed place of business would, in most cases, be obvious.[37] The OECD Commentary, a great number of court decisions and a myriad of scientific contributors defined the contours of the concept quite successfully over the years. The enforceability is intertwined with the presence of physical facilities used by the taxpayer. The tax authorities are able to reach out to the taxpayer with ease and taking enforcement measures to satisfy the non-resident's tax liability, as the need may be, is facilitated to a great extent.[38] Finally, an excessive fragmentation of the non-resident's taxable income between different source countries is limited due to the threshold of the presence of a fixed place of business.[39]

Albeit not manifesting itself as a brick-and-mortar or similar structure, a certain level of physical presence is also required for the purposes of *the dependent agent PE*. For this form of PE to be constituted, the following cumulative conditions must be met: there must be a person who (i) acts in one contracting state on behalf of an enterprise of the other contracting state; (ii) must not constitute an 'independent agent';[40] (iii) must have the authority to conclude contracts in the name of the enterprise or play the principal role leading to the conclusion of contracts that are routinely concluded without material modification by the enterprise;[41] (iv) must exercise this authority in an habitual manner; and (v) must not exercise merely auxiliary and preparatory activities on behalf of the enterprise. The necessity of physical presence in the source country, thus, also exists in the context of dependent agent PE, in the form of a 'person'. Clearly, the person concerned may also be only temporarily present in the country concerned and such presence may not necessarily be linked to one specific place (which can undermine the ability to take enforcement measures to a certain extent when compared to a fixed place of business).[42] Nonetheless, the creation of the nexus is linked to some degree of physical presence.

Finally, *the service PE*, provided as an optional clause by the OECD Commentary[43] and included in article 5(3)(b) of the UN Model, also requires the existence of a physical presence in the source country. As per this last nexus, a PE is considered to be formed in the source country in cases where a non-resident enterprise is

[35] OECD Commentary (2019) ad article 5, § 28.

[36] Hongler and Pistone (2015), p. 23.

[37] Arnold (2003), p. 479.

[38] Arnold (2003), p. 479.

[39] Hongler and Pistone (2015), p. 21.

[40] Under article 5(6) of the OECD MC; Article 5(7) of the UN MC.

[41] Article 12(1) of the MLI; Article 5(5)(a) of the UN MC.

[42] Arnold (2003), p. 479.

[43] OECD Commentary (2019) ad article 5, § 144.

performing services in the country concerned through employees or other personnel who are present in that country for a period or periods exceeding the aggregate 183 days in any twelve-month period. It is important to underline that the texts of the definitions provided by the OECD Commentary and the UN Model differ to a certain extent; and that the OECD Commentary also encompasses an alternative gross revenue threshold. Regardless of these rather minor variations, the requirement of a certain degree of physical presence remains intact.

8.2.2 Concept of Royalties

In a nutshell, royalties can be defined as rental payments received in consideration of the use or the right to use certain intellectual property and information.[44] The more detailed definition provided by article 12(2) of the OECD MC reads as follows: "[. . .] payments of any kind received as a consideration for the use of, or the right to use, any copyright of literary, artistic or scientific work including cinematograph films, any patent, trade mark, design or model, plan, secret formula or process or for information concerning industrial, commercial or scientific experience".[45] Before their deletion in 1992,[46] this definition also contained the words "or for the use of, or the right to use, industrial, commercial or scientific equipment", which are still included in the UN MC.[47] The terms contained in the definition of royalties, such as 'the use of', 'trademark' and 'design', should be interpreted in the light of the contracting states' domestic laws.[48]

The lists provided by article 12(2) of the OECD MC and article 12(3) of the UN MC are intended to be exhaustive.[49] Income generated by the use of other similar rights cannot be considered as royalties, regardless of the domestic laws of the contracting states.[50] It is, however, clear that the contracting states may modify or expand the definition of royalties when they enter into a DTT. This is, in fact, oftentimes the case.[51] For example, some OECD countries extend the definition of royalties to include payments for transmissions by satellite, cable, optic fibre or similar technology.[52]

---

[44] Tadmore (2008), p. 111.

[45] Article 12(2) of the OECD MC.

[46] By the report entitled 'The Revision of the Model Convention' adopted by the OECD Council on 23 July 1992.

[47] Article 12(3) of the UN MC.

[48] Article 3(2) of the OECD MC.

[49] Malherbe et al. (2013), § 54.

[50] Malherbe et al. (2013), § 54.

[51] Holmes (2014), p. 274. A number of countries made reservations on this point [OECD Commentary (2019) ad article 12, § 38–47].

[52] This is, for example, the case of the Hungary-India DTT [Holmes (2014), p. 274].

The payments, which are effectuated for the transfer of the full ownership of an element of property referred to in the royalty definition, cannot be considered as being made in consideration 'for the use of, or the right to use' that property and cannot, therefore, represent a royalty.[53] In fact, where the ownership of rights has been alienated, the consideration cannot be for 'the use of the rights'.[54] Such payments are likely to constitute business profits under article 7 or capital gains within article 13.[55]

Article 12 of the OECD MC grants the exclusive right to tax royalties to the residence state.[56] Source taxation is only allowed in cases where the beneficial owner of the royalties carries on business in the country where the royalties arise through a PE situated therein and the right or property in respect of which the royalties are paid is effectively connected with the PE concerned.[57] In such cases, article 7 becomes applicable. As a consequence, for the purposes of the OECD MC, the threshold requirement in article 12 is effectively the same as the one in article 7 (in conjunction with article 5). Unless a resident of one country has a PE in the other country and the royalties are effectively connected with that PE, the latter country is precluded from taxing the income.[58] The physical presence requirement of the PE is thus transposed to taxation of royalties.

Article 12 of the UN MC, on the other hand, grants a right to tax to the source country up to a certain percentage of the gross amount of royalties, which is established through bilateral negotiations.[59] The UN Model does not provide for any minimum threshold for source country taxation.[60] Royalties are deemed to arise in a contracting state when the payer is a resident of that state.[61] The UN Commentary underlines that, in their bilateral negotiations, the contracting parties may agree upon a general rule that would accept the payer's place of residence as the source of royalty, and an exception, which provides that where the right or property for which the royalty was paid was used in the country having a place-of-use rule, the royalty would be deemed to arise in the country concerned.[62] This additional regulation may reveal to be an effective remedy to cases where the location of the payer of royalties is dissociated from the place of use.

Just as under the OECD MC, the UN MC provides that in cases where the beneficial owner of the royalties carries on business in the country in which the royalties arise through a PE situated therein, and the right or property in respect of

[53] OECD Commentary (2019) ad article 12, § 8.2.

[54] OECD Commentary (2019) ad article 12, § 8.2.

[55] OECD Commentary (2019) ad article 12, § 8.2.

[56] Article 12(1) of the OECD MC.

[57] Article 12(3) of the OECD MC.

[58] Arnold (2003), p. 481.

[59] Article 12(2) of the UN MC.

[60] Arnold (2003), p. 481.

[61] Article 12(5) of the UN MC.

[62] UN Commentary (2021) ad article 12, § 28.

which the royalties are paid is effectively connected to the PE concerned, article 7 (in conjunction with article 5) becomes applicable instead of article 12.[63] Under such a scenario, the physical presence requirement of the PE is transposed to taxation of royalties.

Despite the lack of a source taxation allowed under the OECD MC, DTTs granting a right to tax royalties to source countries are not rare in practice. Unless the royalties are effectively connected to a PE situated in the countries concerned, this particular category of income can be submitted to source taxation without any minimum threshold.

To the extent where a business profit can be qualified as a royalty under article 12, source taxation can thus be effectuated without the need of a physical presence. As a matter of fact, the inclusion of payments for the use of industrial, commercial and scientific equipment within the royalty definition in article 12 of the 1963 Draft of the OECD MC[64] was directly linked with the possibility for companies in the leasing sector to be established in countries where they could not be submitted to tax due to the lack of a PE, despite the fact that the leased equipment was being used in the country concerned.[65] The inclusion of equipment leasing in the royalty definition, which gave rise to the levy of a withholding tax in source countries, was perceived to be the most appropriate means to combat base erosion occurring due to the mobile nature of leasing activities, which did not necessitate local infrastructure to be carried on in a country.[66] As demonstrated by this historical example, article 12 may reveal to be a useful tool of source taxation in cases where the PE threshold cannot be met.

8.2.3 Concept of Technical Services Fees

Article 12A, entitled 'fees for technical services', was added to the UN MC in 2017.[67] The OECD MC does not contain a comparable article. In fact, the term 'technical services fees' is mentioned neither in the OECD MC nor in the OECD Commentary.[68] Technical services fees constitute a special category of income that was added to the UN MC to eliminate the ambiguity caused by the rather wide use of the term in DTTs in a 'not-always-coherent' manner.

A non-negligible number of developing countries have been inserting this additional source of income in their DTTs for a certain period of time. While the concept

[63] Article 12(4) of the UN MC.

[64] As already mentioned, these terms were deleted from the OECD MC in 1992.

[65] Jiménez (2018), p. 631.

[66] Jiménez (2018), p. 631.

[67] UN Commentary (2021) ad article 12A, § 1.

[68] The OECD Commentary does, however, refer to 'provision of services' and 'technical assistance' [OECD Commentary (2019) ad article 12, § 11.1–11.6].

is being used in international tax law, an international consensus on its scope is lacking.[69] Some countries consider that provision of 'brain-work' and technical services are already covered by the concept of royalties,[70] while some other countries opt for a brief definition or, at least, an explicit mention of technical services fees in their DTTs.[71] Not surprisingly, the scope of the term 'technical services fees' varies to a great extent under the domestic laws of the countries as well.[72] In some countries technical services refer to services that involve the application of specialist knowledge such as know-how and engineering expertise;[73] in some others they relate to services connected with the transfer of knowledge or expertise; [74] and finally some countries opt for a wide approach and take the view that almost all of the services provided to the country's residents by non-resident providers can be qualified as technical services.[75]

Article 12A(3) of the UN MC defines the concept, in an exhaustive manner,[76] as "[. . .] any payment in consideration for any service of a managerial, technical or consultancy nature [...]".[77] Are excluded from the scope of the term, payments made: to an employee of the person making the payment, for teaching in/by an educational institution and by an individual for services for the personal use of an individual.[78]

The UN Commentary aims to contribute to determining the contours of the concept of 'technical services fees'. The Commentary indicates that the services must involve the application by the service provider of specialised knowledge, skill or expertise on behalf of a client or the transfer of knowledge, skill or expertise to a client[79] (other than a transfer of information covered by the definition of 'royalties'[80]).[81] Consequently, services of a routine nature that do not involve the application of such specialised knowledge, skill or expertise cannot be considered as 'technical services'.[82] This interpretation seems to be in line with the view adopted

[69] See van der Bruggen (2001), pp. 42–43.

[70] UN Commentary (2021) ad article 12A, § 5.

[71] See van der Bruggen (2001), pp. 42–43.

[72] Pickering (2012), p. 33.

[73] Pickering (2012), p. 33.

[74] Pickering (2012), p. 33.

[75] This is, for example, the case in Brazil [Rocha (2012), p. 158].

[76] UN Commentary (2021) ad article 12A, § 61.

[77] It is interesting to note that an almost identical definition was already contained in bilateral treaties of India, Pakistan and Malaysia [van der Bruggen (2001), pp. 42–43].

[78] Article 12A(3) of the UN MC.

[79] In the same vein, the OECD Commentary defines contracts for 'provision of services' in this context as contracts in which one of the parties undertakes the use of the customary skills of their calling to execute the work themself for the other party [OECD Commentary (2019) ad article 12, § 11.2].

[80] For the difference between transfer of 'know-how' and 'technical services fees', see, for example, van der Bruggen (2001), pp. 50–52; and UN Commentary (2021) ad article 12A, § 85 and 100–103.

[81] UN Commentary (2021) ad article 12A, § 62.

[82] UN Commentary (2021) ad article 12A, § 62.

by Indian Courts, which indicated in a number of different decisions that a mere collection of a fee for use of a standard facility or equipment (which may have been developed by the enterprise commercialising its use), without any human intervention, does not amount to provision of technical services.[83] The UN Commentary completes its definition by describing the terms 'management', 'technical' and 'consultancy' and underlining that these three terms, which do not have precise meanings, may overlap.[84]

Article 12A of the UN MC grants a right to tax to the source country up to a certain percentage of the gross amount of technical services fees, which is established through bilateral negotiations.[85] Source taxation provided by article 12A does not require any threshold, such as a PE or a minimum period of presence in the source country.[86] This lack of threshold is consistent with the rationale behind the existence of this allocation rule: to be able to submit to tax the income obtained by a non-resident from a country with little or no presence in the country concerned, combined with an aim to circumvent any base-erosion and profit-shifting.[87]

The technical services fees are deemed to arise in the residence country of the payer.[88] In their bilateral negotiations, the contracting parties may agree upon a general rule, which would accept the payer's residence as the source of fees for technical services fees, suffering from one exception, which would deem that payments arise in the contracting state (having a place-of-use rule) where the technical services are used or consumed.[89]

As is the case with royalties, if the beneficial owner of technical services fees carries on business in the country in which the fees for technical services arise through a PE and the fees for technical services are effectively connected with such PE, article 7 (in conjunction with article 5) becomes applicable instead of article 12A.[90] Under such a scenario, the requirement of physical presence is, thus, transposed to source taxation of technical services fees.

8.2.4 Concept of Income from Automated Digital Services

'Income from automated digital services' is a new category of income, added in the UN MC in 2021 to preserve the taxing rights established under the domestic laws of source countries.[91] Article 12B, regulating income from automated digital services,

[83] See Mehta (2016), p. 2.
[84] UN Commentary (2021) ad article 12A, § 63–67.
[85] Article 12A(2) of the UN MC.
[86] UN Commentary (2021) ad article 12A, § 33.
[87] UN Commentary (2021) ad article 12A, § 33.
[88] Article 12A(5) of the UN MC.
[89] UN Commentary (2021) ad article 12A, § 123.
[90] Article 12A(4) of the UN MC.
[91] UN Commentary (2021) ad article 12B, § 2.

forms, as a matter of fact, the solution found by the UN to the problems arising in taxation of the digital economy.

Article 12B(5) defines the term 'automated digital services' as "[...] any service provided on the Internet or another electronic network, in either case requiring minimal human involvement from the service provider". The UN Commentary specifies that a service is considered to be automated "[...] when the user is able to make use of the service because of equipment and systems being in place, which allow the user to obtain the service automatically, as opposed to requiring a bespoke interaction with the supplier to provide the service".[92] The human involvement necessary to create, support or maintain the system needed for the provision of services, or making other generic adjustments that are unrelated to individual user requests, does not constitute a human involvement within the meaning of the definition of the term.[93] The same holds true for human intervention where the provision of services to new users involves very limited human response to individual user requests/input.[94] The purpose is to cover the services in relation to which enterprises can scale up and provide the same type of services to new users with minimal human involvement.[95]

The concept of automated digital services does not distinguish between different Internet or electronic network transmission methods. Whether or not the service provider owns, leases or otherwise controls the transmission equipment is not relevant either.[96]

The definition provided by article 12B(5) is exhaustive.[97] In its paragraph 6, the article contains a list of specific services that will often constitute automated digital services. The list concerned[98] is only an exemplative list,[99] akin to the list contained in article 5(2) of both Model Conventions. To be qualified as an automated digital service, a service must fall within the definition contained in paragraph 5.[100] Therefore, a service specifically mentioned in paragraph 6 may not necessarily constitute an automated digital service. In the same vein, a number of services that are not mentioned in paragraph 6 may form automated digital services. To avoid uncertainty to a certain extent, the UN Commentary also provides a list of activities that are not covered by the concept of 'automated digital services'. The activities concerned are customised professional services, customised online teaching

[92] UN Commentary (2021) ad article 12B, § 53.

[93] UN Commentary (2021) ad article 12B, § 53.

[94] UN Commentary (2021) ad article 12B, § 53.

[95] UN Commentary (2021) ad article 12B, § 54.

[96] UN Commentary (2021) ad article 12B, § 55.

[97] UN Commentary (2021) ad article 12B, § 56.

[98] The list is composed of online advertising services, supply of user data, online search engines, online intermediation platform services, social media platforms, digital content services, online gaming, cloud computing services and standardised online teaching services.

[99] UN Commentary (2021) ad article 12B, § 57.

[100] UN Commentary (2021) ad article 12B, § 57.

services, services providing access to the Internet or to another electronic network, online sale of goods and services other than automated digital services and revenue from the sale of a physical good, irrespective of network connectivity (IoT).[101]

As elucidated by its paragraph 7, article 12B does not apply to any amounts falling within the definition of royalties in the sense of article 12 or of fees for technical services within the meaning of article 12A. Given the nature of the services, it cannot be excluded that a part of a bundle of automated digital services will be more akin to royalties and/or technical services fees and the remaining part will consist of automated digital services. In such cases, either a severance may be effectuated (and thereby the income referring to each component may be submitted to tax under the relevant allocation rule) or, in cases where a severance is not feasible, the overall service may be covered by one of the allocation rules, which is predominant for the purposes of the bundle of services concerned.[102]

Article 12B grants a right to tax to the source country up to a certain percentage of the gross amount of the payments underlying the income from automated digital services.[103] While the rate is to be bilaterally negotiated, the UN Commentary suggests a modest rate, such as 3 or 4%, given the possibility of double or excessive taxation.[104] An option to pay tax on a net profit basis is also provided to the taxpayer in paragraph 3 of the article.[105] Article 12B does not require any particular threshold, such as a PE or a minimum period of presence in a source country.[106] This absence of threshold is in line with the purpose of this new allocation rule, which is to submit to tax income derived from automated digital services rendered by non-resident enterprises with little or no physical presence in the source country.[107]

Income from automated digital services is deemed to arise in the country where the underlying payments for the income concerned are made by a resident of the relevant country or, alternatively, in the country where the person making the underlying payments for automated digital services has a PE in connection with which the obligation to make the payments was incurred, and such payments are borne by the PE.[108] On this point, the UN Commentary indicates that where there is an obvious economic link between the automated digital services provided and the PE of the payer to which the services are provided, the income from automated digital services is considered to arise in the country in which the PE is situated.[109] In cases where there is no economic link between the automated digital services and the PE, the income from automated digital services is considered to arise in the country

[101] UN Commentary (2021) ad article 12B, § 59.

[102] UN Commentary (2021) ad article 12B, § 61–62.

[103] Article 12B(2) of the UN MC.

[104] UN Commentary (2021) ad article 12B, § 4.

[105] Article 12B(3) of the UN MC; UN Commentary (2021) ad article 12B, § 3.

[106] UN Commentary (2021) ad article 12B, § 7.

[107] UN Commentary (2021) ad article 12B, § 7.

[108] Article 12B(9) of the UN MC.

[109] UN Commentary (2021) ad article 12B, § 72.

of which the payer of the consideration for automated digital services is a resident.[110]

The UN Commentary emphasises that to the extent where the payer of the consideration for automated digital services is not a resident of either contracting states, article 12B does not apply to the income derived from automated digital services unless the payer has a PE in the contracting state and a clear link between the automated digital services and the PE exists.[111] Otherwise, a force-of-attraction principle for payments in consideration for automated digital services would be put in place and such a regulation would reveal to be inconsistent with other provisions of the UN MC.[112]

The sourcing rule contained in paragraph 9 of the article operates on the basis of payment and not on the basis of the location of users. In fact, the UN Commentary underlines that no taxing rights are allocated to the country where users are employing automated digital services for free under article 12B.[113] As per the Commentary, the voluntary or involuntary provision of data by users as a condition to access a certain digital service should not be classified as a type of payment made in consideration for the automated digital service.[114] The effect of the attribution effectuated on the basis of the payment can be explained by the following example in the context of a multi-sided platform: a resident of State C employs the automated digital services rendered by an enterprise located in State A to reach the user base of State B. The payment made by the resident of State C to the enterprise situated in State A may be submitted to tax in State C under article 12B of the DTT concluded between the two countries. State B, however, is not entitled to any taxing rights on the mere basis of being the country in which users are located.[115]

As is the case with royalties and technical services fees, if the beneficial owner of the income from automated digital services carries on business in the country in which the income from such services arises through a PE and the services concerned are effectively connected with such PE, article 7 (in conjunction with article 5) becomes applicable instead of article 12B.[116] Accordingly, under such a scenario, the requirement of physical presence is transposed to source taxation of income from automated digital services.

[110] UN Commentary (2021) ad article 12B, § 73.

[111] UN Commentary (2021) ad article 12B, § 73.

[112] UN Commentary (2021) ad article 12B, § 73.

[113] UN Commentary (2021) ad article 12B, § 75.

[114] UN Commentary (2021) ad article 12B, § 74.

[115] UN Commentary (2021) ad article 12B, § 75.

[116] Article 12B(8) of the UN MC.

8.3 Permanent Establishment: The Fairest Nexus of All

> [. . .] [I]t is not clear why a single employee or a single warehouse should create a substantial nexus while "physical" aspects of pervasive modern technology should not.[117]

The claim of a right to tax a portion of the profits obtained by companies offering highly digitalised services often made by countries where users are situated should not be interpreted as trying to go beyond the consensus reached in the 1920s and still reigning the international tax law or to aim for ring-fencing the digital economy.

In the quest for an efficient solution to tackle the 'lack of source taxation problem', the authors agree with the following observation made by Jiménez, "[. . .] before throwing away what has been consolidated over many years of evolution, one must be sure that something new is indeed needed and would be better than the old tools, even if the old tools may need to evolve rather than being changed in a, more or less, radical form".[118] Jiménez also argues that it is difficult to understand why the contributors in favour of more 'market' taxation do not more actively promote traditional standards for the taxation of business profits that are already defined, mainly, in the context of the UN MC (in its post-2017 version), linked with a less formalistic reading of article 5 of the OECD MC and the UN MC, as a means to enforce and tax relevant business presence, digital or otherwise, in a country.[119] The authors also agree with this assertion, albeit only to a certain extent.

In our opinion, in establishing a simple, efficient and viable solution, one should indeed first attempt to use the 'old tools' at their disposal to the extent possible. In the context of the digital economy, however, the old tools do not provide for a solution, as explained above, at least in their current wording. While the contours of the rules contained in DTTs may be expanded to a certain extent by means of interpretation in view of taking into consideration new economic realities or technological developments, it should not be forgotten that the interpretation method has its limits. Such method cannot be relied upon when the possible meaning of the text used by a DTT cannot cover the developments occurred in real-life.[120] For the purposes of the digital economy, the emergence of new factors, diverging to a great extent from circumstances envisioned when the texts of the Model Conventions were prepared, would render the mere employment of the interpretation method to cover the recent concepts inappropriate.[121] A PE concept, that is clearly necessitating the existence of a physical presence, cannot—and should not—be stretched as far as to cover activities that do not involve any form of physical presence. For instance, it is not possible to construe by means of interpretation that data are covered by the

[117] *South Dakota, Petitioner, v. Wayfair Inc., et al.*; Supreme Court of the United States, No. 17-494, 21 June 2018, Opinion of the Court, p. 15.

[118] Jiménez (2018), p. 620.

[119] Jiménez (2018), p. 637.

[120] In the same opinion, Zhu (2014), p. 8.

[121] Zhu (2014), p. 8.

exemplative lists of articles 5(2)(f) of the OECD and UN Model Conventions, which mention "a mine, an oil or gas well, a quarry or any other *place of extraction of natural resources*"[122] as suggested by Petruzzi and Buriak,[123] despite a number of similar characteristics shared by data and natural resources underlined throughout this book. Without any doubt, to become applicable in the context of the digital economy, the PE concept needs to be modified.

Building on these primary observations, this section concentrates on finding a viable solution to the systematic lack of source taxation in the context of the digital economy. It is first argued that the current system results in ring-fencing the traditional economy. It is then asserted that the tendency to provide excessive shielding to the digital economy does not constitute a reasonable long-term strategy. At some point the need to go beyond the physical borders of the PE concept would need to be satisfied. Following this reflection, a new virtual PE nexus is suggested. This section concludes by the proposal to incorporate an article regulating 'digital services fees' in DTTs, as a residual category to complement the virtual PE. As it can be observed, the 'new solutions' suggested are built on modified 'old tools'.

This analysis is conducted in accordance with the main line of reasoning adopted in this book. Consequently, this section exclusively concentrates on the examination of multi-sided platforms subsidising users to provide digital services to the customers of the other side(s). Beyond any doubt, a similar approach could be applied to the broader digital economy, subject to further analysis. The majority of the work cited in this section concentrates on the digital economy, and not on the multi-sided platforms, to begin with. That being said, the authors did not effectuate their analysis with the overall digital economy in mind and opted for the narrow approach of exclusively examining multi-sided platforms.

As a final remark, it is important to underline that as demonstrated in the last decade, implementation of new measures by means of a multilateral instrument would constitute a more efficient option. While such a method would be recommended, it is not clear whether it could effectively be implemented for the purposes of the taxation of multi-sided platforms. Subsequently, some remarks on bilateral negotiations and implementation of the newly suggested provisions in the DTTs are also made throughout this section.

8.3.1 'Ring-Fencing' the 'Traditional' Economy

Establishing a new nexus to cover some services that are provided by highly digitalised enterprises, which cannot be currently covered by any existing nexus, inevitably gives rise to the question of whether such a new and specific nexus would

[122] Emphasis added by the authors.

[123] Petruzzi and Buriak (2018), p. 17.

violate the principle of neutrality. In a nutshell, the principle of neutrality, which is one of the key components of the Ottawa Taxation Framework Conditions, dictates that taxation should seek to be neutral and equitable between forms of business activities, which entails that the same principles of taxation should apply to all forms of business.[124] On top of observing the principle of neutrality, adoption of a new nexus should not result in a lack of efficiency either. Efficiency, which is again one of the key components of the Ottawa Taxation Framework Conditions, can be described as the minimisation of compliance costs to businesses and administration costs for governments.[125] In the light of these two key principles, among others (such as certainty and simplicity), the new nexus should not result in ring-fencing enterprises providing highly digitalised services.

The issue of 'ring-fencing the digital economy' has been discussed on a myriad of occasions throughout the digital economy debate. The OECD has shown a great deal of reticence to implement any measures specifically directed at companies providing highly digitalised services with an aim of not 'ring-fencing the digital economy'. This restraint can also be observed in the end results of the OECD's work in the area, namely the relevant dispositions of the MLI as well as the Two-Pillar Approach.

While it is clear that enterprises providing digital services should not be ring-fenced, the current climate of 'over precaution' does not seem to be justified either. Application of specific rules, diverging from the general taxing principles and standards applicable to the whole economy to a certain extent, is not a recent phenomenon that has been put forward exclusively within the context of the so-called 'digital economy'. General principles of tax law should be, and were on several occasions, broken down in order to take into consideration the real-life economic specificities. For instance, entertainers and sportspersons have been subject to different rules from other employees and self-employed persons under both the OECD and the UN Models.[126] Given the long-lasting existence of these specific rules, the artificial separation effectuated for entertainers and sportspersons seems to have been widely accepted and not interpreted as being in contradiction with the international tax law system.[127] In the same vein, and with a closer relation to the issue at hand, the dependent agent PE currently contained in article 5(5) of both Model Conventions was enacted to allow source taxation in cases where no fixed place of business existed in the country concerned. The dependent agent rule differs to a great extent from the fixed place of business threshold.[128] While the latter focuses on the existence of a fixed place, the former concentrates on the activities of the taxpayer carried on by dependent agents.[129] As demonstrated by the inclusion of this divergent threshold into the Model Conventions, despite the frequently

[124] OECD (2015), § 10.

[125] OECD (2015), § 10.

[126] Article 17 of the OECD MC; Article 17 of the UN MC.

[127] Schön (2018), p. 281.

[128] Arnold (2003), p. 479.

[129] Arnold (2003), p. 479.

expressed opinion of the need for adapting the 'outdated' nineteenth century PE concept to the economy of the twenty-first century, both Model Conventions have recognised for a substantially long time that the 'fixed place of business' part of the PE definition was not appropriate for all business activities and that a different threshold was necessary for some of them.[130]

As a matter of fact, and as opposed to the widespread narrative, it could be argued that in its current wording article 5 of the Model Conventions is ring-fencing the 'traditional' economy. The current fixed place PE definition only affects enterprises operating abroad through a fixed place of business.[131] Similarly, with regard to dependent agent PE, only enterprises operating through an agency structure are affected.[132] The current situation is, therefore, "boosting the digital economy with considerable and perfectly legitimate tax savings".[133]

On this very point, it is important to underline that the existence of VAT/GST or other similar consumption taxes cannot be interpreted as amounting to 'rebalancing' the source country taxation with respect to digital services. As underlined by Schön, a global rise of general consumption taxes occurred simultaneously to the international compromise on business taxation reached in the 1920s.[134] The predominant role of VAT/GST as a source of revenue for destination countries cannot be subject to debate.[135] This being stated, the scope of a VAT/GST or a similar general consumption tax extends to the majority of available goods and services on the market by the very definition of such taxes. In other words, the goods and services commercialised by enterprises active in the 'traditional' economy are subject to the general consumption taxes applicable in a certain country, just as the digital services provided by highly digitalised enterprises. Given their neutral nature, as it should be, general consumption taxes do not distinguish between goods and services provided by 'highly digitalised' and 'traditional' enterprises. As a natural consequence, the existence of such taxes cannot be interpreted as counterbalancing the current ring-fencing of the traditional economy from a corporate income tax perspective.

Needless to remind, the ring-fencing of the traditional economy is occurring as an additional layer on top of the fact that enterprises active in this type of economy are actually paying for their raw material.

Before concluding, making an observation with regard to compliance costs to businesses (in relation to the principle of efficiency) is in order. As expressed by the Supreme Court of the US in the South Dakota v. Wayfair, Inc. decision, "[...] the administrative costs of compliance, especially in the modern economy with its Internet technology, are largely unrelated to whether a company happens to have a physical presence in a State"; "[...] a small company with diverse physical presence

[130] Arnold (2003), p. 479.
[131] Hongler and Pistone (2015), p. 42.
[132] Hongler and Pistone (2015), p. 42.
[133] Brauner and Pistone (2017), p. 3.
[134] Schön (2018), p. 286.
[135] Schön (2018), p. 286.

might be equally or more burdened by compliance costs than a large remote seller[,] [t]he physical presence rule is a poor proxy for the compliance costs faced by companies that do business in multiple States".[136]

8.3.2 *Going Beyond 'Physical' Borders, the Virtual PE*

8.3.2.1 Rationale Behind the Need of Going Beyond Physical Borders

The concept of PE should be considered as a compromise that aims to strike a fair balance between the exclusive taxation of the residence state of the enterprise and the right of the source country to have a fair share of taxes in respect of business activities that are stably located on its territory.[137] As a natural result of being a tool implementing a compromise between resident states and source states, the concept of PE should not amount to the predominant empowerment of one category of state over the other. While safeguarding the interests of residence countries is necessary, the application of the concept should not systematically result in an excessive limitation of the source countries' right to tax. In the context of the digital economy, the concept of PE in its current form is acting as a serious limitation on the ability of source countries to submit to tax the activities taking place in their territory, privileging, thereby, residence countries. There exists, thus, a need of 'rebalancing' the well-established and the well-functioning concept of PE, to adapt it to the new economic circumstances.

What is preventing a PE to be formed in the context of digital services rendered remotely is the requirement of a certain degree of physical presence. While it has been argued for a rather long time that the necessity of a physical presence in the source country does not constitute the most appropriate threshold to submit highly digitalised enterprises to tax, the OECD, in line with the 1998 Ottawa Declaration, has opted for retaining the requirement concerned as a condition for admitting the existence of a PE.[138] Undisputedly, with each passing year, the physical presence rule becomes further removed from the economic reality.[139] This persistence on physical presence in spite of the utter reality that a substantial amount of business can be conducted without any need for physical presence within a country in the 'modern' economy,[140] turned the concept of PE into a 'cage' that prevents source

[136] *South Dakota, Petitioner, v. Wayfair Inc., et al.*; Supreme Court of the United States, No. 17-494, 21 June 2018, Opinion of the Court, p. 12.

[137] Hongler and Pistone (2015), p. 15.

[138] Brauner and Pistone (2017), p. 1.

[139] *South Dakota, Petitioner, v. Wayfair Inc., et al.*; Supreme Court of the United States, No. 17-494, 21 June 2018, Opinion of the Court, p. 10.

[140] *Quill Corporation, Petitioner, v. North Dakota By and Through its Tax Commissioner, Heidi HEITKAMP*, 504 U.S. 298, 26 May 1992, § 18–19.

countries from exercising their taxing powers.[141] It can, therefore, be asserted that the physical presence requirement turned into physical borders, which limit the applicability of the concept of PE to the 'traditional' (or the so-called 'brick-and-mortar') economy. It is high time to go beyond those physical borders.

From a theoretical standpoint, the creation of a new PE nexus is justified on the basis of the benefit theory.[142] As argued by Pinto, as early as in 2003, the fact that an enterprise does not have a physical presence in the source country does not necessarily entail that the enterprise concerned cannot benefit from the infrastructure put in place by that country.[143] Enterprises providing highly digitalised services rely, indeed, on a number of contributions made by the source country. These contributions can be regrouped in four main categories:[144]

1. the contribution to the business environment and economy; including, among others, the general business confidence, measures taken against corruption as well as affluence and ability to consume;
2. the contribution to the technological infrastructure; encompassing, for instance, suitable telecommunications infrastructure, Wi-Fi and broadband;[145]
3. the contribution to the legal system; which entails the existence and the maintenance of an efficient legal system that would allow, among others, enforcement of customers' payments, protection of intellectual property rights as well as maintenance of a competitive and conducive business environment;
4. the contribution to infrastructure; comprising of many factors, such as a relevant level of energy supply, an appropriate mechanism of waste recycling and a general modernised infrastructure.

The benefits offered by a source country in the context of the digital economy have been eloquently described by the Supreme Court of the US in its Wayfair decision. While examining whether Wayfair, Inc., an online retailer of home goods and furniture, should be collecting sales tax in a State where it does not have a physical presence, the Court employed the benefit theory and made the following observations: "What Wayfair ignores [. . .] is that creating a dream home assumes solvent state and local governments. State taxes fund the police and fire departments that protect the homes containing their customers' furniture and ensure goods are safely delivered; maintain the public roads and municipal services that allow communication with access to customers; support the 'sound local banking institutions to support credit transactions [and] courts to ensure collection of the purchase price'

[141] Brauner and Pistone (2017), p. 1.

[142] For a thorough theoretical analysis, see Hongler and Pistone (2015), pp. 15–22.

[143] Pinto (2003), pp. 22–23.

[144] Elliffe (2021), pp. 33–34. See also Hongler and Pistone (2015), p. 22; and Pinto (2003), p. 22.

[145] Maintaining a telecommunications structure apt to cope with the current data traffic incurs substantial costs as demonstrated by the recent debate in the EU. For more information see, for example, Euronews (2022).

[...]; and help create the 'climate of consumer confidence' that facilitates sales [...]".[146]

The levy of a data collection tax in *step one* should not be interpreted as breaking the link between the contributions made by a source country and the digital services rendered by a multi-sided platform. First of all, in reality, the data collection tax is to be levied by countries for practical purposes,[147] preferably as an interim measure until a well-functioning and efficient data labour market, allowing a fair and direct remuneration of the users in exchange of their data, can be implemented. In other words, the tax is levied as a collective compensation for the purchase of raw material in the name of all users located in a country, until a personal remuneration mechanism can be put in place. Thus, the data collection tax does not constitute a compensation for the benefits provided by source countries to multi-sided platforms undertaking business activities in relation to their territories. Are amongst such benefits, a well-functioning economy, consumers with appropriate purchasing power and a certain degree of digital literacy, adequate technological infrastructure and an effective legal system.

8.3.2.2 New Nexus, Not a New Suggestion

Stating that despite not being physically present, multi-sided platforms clearly have a 'substantial nexus' in source countries would not constitute a far-reaching argument. The term 'substantial nexus' was used in a number of court decisions rendered in the US. In its Wayfair decision, albeit not precisely defining the concept, the Supreme Court of the US states that "[s]uch a nexus is established when the taxpayer [or collector] 'avails itself to the substantial privilege of carrying on business' in that jurisdiction".[148] In examining the Act of South Dakota imposing an obligation to collect sales taxes to enterprises, which deliver more than USD 100,000 of goods or services into South Dakota or engage in 200 or more separate transactions for the delivery of goods and services into the State on an annual basis, the Court observed that "[t]his quantity of business could not have occurred unless the seller availed itself of the substantial privilege of carrying business in South Dakota".[149]

---

[146] *South Dakota, Petitioner, v. Wayfair Inc., et al.*; Supreme Court of the United States, No. 17-494, 21 June 2018, Opinion of the Court, pp. 16–17.

[147] See Chap. 6, entitled 'Ending the era of free data: data collection tax', for more details on this point.

[148] *Polar Tankers, Inc. v. City of Valdez*, 557 U.S. 1, 15 June 2009 cited in *South Dakota, Petitioner, v. Wayfair Inc., et al.*; Supreme Court of the United States, No. 17-494, 21 June 2018, Opinion of the Court, p. 22.

[149] *South Dakota, Petitioner, v. Wayfair Inc., et al.*; Supreme Court of the United States, No. 17-494, 21 June 2018, Opinion of the Court, p. 23.

A new PE nexus based on the concept of significant economic presence was suggested in 2015 by the OECD in the BEPS Action 1[150] Final Report.[151] In 2018, the European Commission took that suggestion one step further and published a proposal for a 'Council Directive laying down rules relating to the corporate taxation of a significant digital presence' (the Proposed Directive),[152] making, thereby, the first genuine attempt to adapt the PE rule to the realities of the twenty-first-century.[153] In a nutshell, the Proposed Directive aims to establish a new PE nexus based on significant digital presence and to determine the relevant rules for attribution of profits. Under article 4(1) of the Proposed Directive, a PE shall be taken to exist for the purposes of corporate income tax, to the extent where a significant digital presence exists through which a business is wholly or partly carried on. The proposal establishes the existence of a significant digital presence on the basis of the following *alternative* thresholds: the proportion of total revenues[154] obtained in a given tax period and resulting from the supply of digital services through a digital interface[155] to users located in a Member State in that tax period exceeds EUR 7,000,000; or the number of users of one or more of the digital services concerned who are located in a Member State exceeds 100,000 in that tax period; or the number of business contracts for the supply of any such digital service that are concluded in that tax period by users located in a Member State exceeds 3000.[156]

In a similar vein Hongler and Pistone suggest incorporating a new PE nexus, in the form of a new paragraph, into article 5 of the OECD MC. The nexus definition put forward by the contributors is as follows: "If an enterprise resident in one Contracting State provides access to (or offers) an electronic application, database, online marketplace, storage room or offers advertising services on a website or in an electronic application used by more than 1000 individual users per month domiciled in the other Contracting State, such enterprise shall be deemed to have a permanent establishment in the other Contracting State if the total amount of revenue of the enterprise due to the aforementioned services in the other Contracting State exceeds XXX (EUR, USD, GBP, CNY, CHF, etc.) per annum".[157]

The contributors underline a number of elements in relation to this new definition. First, the new nexus should define the actual revenue grounds that shall be affected

---

[150] Tax Challenges Arising from Digitalisation, https://www.oecd.org/tax/beps/beps-actions/action1/. Accessed 7 July 2023.

[151] OECD (2015), § 277–291.

[152] European Commission (2018).

[153] On this point, see Escribano (2019).

[154] 'The proportion of total revenues' is to be determined in proportion to the number of times that devices are used in that tax period by users located anywhere in the world to access the digital interface through which the digital services are supplied [article 4(7) of the Proposed Directive].

[155] 'Digital interface' is defined as any software, including a website or a part thereof and applications, including mobile applications, accessible by users in article 3(2) of the Proposed Directive.

[156] Article 4(3) of the Proposed Directive.

[157] Hongler and Pistone (2015), p. 25.

by the new provision.[158] Second, the contributors argue that a user-based, and not a customer-based, threshold should be established to reflect better the intensity of the significant digital presence.[159] On this point, the contributors also note that the amount of time spent by users on a specific online platform would constitute an even better threshold.[160] Third, inserting a time threshold could be reflected upon, especially when the detrimental consequences of non-compliance in certain jurisdictions along with the fast-growing potential of certain online businesses are taken into consideration.[161] For instance, the PE may be formed 12 months after the enterprise breached the user threshold.[162] Fourth and last, instead of including a fixed amount of revenue threshold, reference can be made to a determined ratio relating to the overall wealth or economic power of a jurisdiction.[163]

8.3.2.3 Virtual PE

In line with the suggestions already made by, among others, the OECD, the European Commission as well as Hongler and Pistone, the authors assert that the best manner to tackle the current source taxation issues posed by the business model of multi-sided platforms would be to establish a virtual PE nexus. As also argued by Avi-Yonah, "[i]n the medium to long run, [...] a virtual PE solution is inevitable".[164] Delaying the inevitable by trying to concentrate on other issues and potential solutions is far from constituting an efficient problem solving method. Such an approach could only cause more complexities.

The purpose of this section is to elaborate the relevant features that need to be present in the definition of the new nexus or that need to be considered when such definition is being prepared; namely activities covered, revenue threshold, user threshold and business contract threshold. A draft version of the definition is not provided by the authors for one simple reason; most of the elements mentioned below require a thorough analysis of the economic and social factors, which are beyond the expertise of the authors. Establishing a virtual PE nexus constitutes a major task, which needs to be effectuated by comprehensive inter-disciplinary scientific research.

[158] Hongler and Pistone (2015), p. 24.
[159] Hongler and Pistone (2015), p. 25.
[160] Hongler and Pistone (2015), p. 25.
[161] Hongler and Pistone (2015), p. 25.
[162] Hongler and Pistone (2015), p. 25.
[163] Hongler and Pistone (2015), p. 25.
[164] Avi-Yonah (2018), p. 164.

8.3.2.3.1 Activities Covered

In our view, providing *a broad definition* of the activities covered, as effectuated by the Proposed Directive,[165] would be a better option than enumerating the actual revenue grounds that shall be affected by the new nexus. Technology is evolving at an ever-accelerating pace and new 'virtual realities' make their entrance into our daily lives with a tremendous speed. It is almost impossible to foresee the transactions that might be conducted by means of virtual platforms and the types of revenues that may be obtained on that basis. To state the obvious, how could have one predicted with certainty the substantial revenues that may be obtained by means of non-fungible tokens (NFTs) or through metaverse? Opting for a broad definition would eliminate the need to update the list of activities regularly, an exercise which would, beyond any doubt, reveal to be extremely cumbersome.

8.3.2.3.2 Revenue Threshold

In our opinion, a revenue threshold should not be included in the definition of the virtual PE. The primary aim of such threshold seems to be preventing the excessive fragmentation of highly digitalised enterprises' profits between a substantial number of source countries.[166] When the nexus is established on the basis of objective criteria, such as a fixed place, and the business activities conducted in relation to a country demonstrate a certain level of intensity, a revenue threshold would not be relevant, since the attribution of profits to a source country where a 'significant presence' exists cannot be interpreted as amounting to excessive fragmentation. This view is supported by the fact that a revenue threshold does not exist for other types of PE, which require the existence of a physical presence in the source country. The key issue is thus to properly establish the significant digital presence of multi-sided platforms in source countries. Such establishment can be effectuated, cumulatively or alternatively, by means of a user threshold and a business contract threshold.[167] Some explanations regarding a revenue threshold are nonetheless made, given the fact that such a threshold may need to be incorporated as a part of a political compromise.

The revenues taken into consideration in the context of the threshold should exclusively be the revenues obtained from the activities falling within the scope of the virtual PE nexus and not revenues deriving from other business activities of highly digitalised enterprises. A revenue threshold may take the form of a fixed amount or a determined ratio relating to the overall wealth or economic power of a jurisdiction.[168] Implementing a fixed amount revenue threshold may, *prima facie*,

[165] See article 4(3) of the Proposed Directive.

[166] Hongler and Pistone (2015), p. 25.

[167] As it is the case in the Proposed Directive, see article 4(3)(b) and (c) of the Proposed Directive.

[168] Hongler and Pistone (2015), p. 25.

seem more suitable, especially due to the level of certainty it may add to the virtual PE definition.[169] While alleviating a number of practical difficulties that may arise in the context of a fixed amount threshold, a determined ratio threshold provides less certainty and gives rise to additional implementation challenges.

From a taxpayer perspective, basing the monetary threshold on gross revenue would be advisable. As a matter of fact, an income-based threshold for business profits would require taxpayers to calculate net profits in order to determine whether the threshold is satisfied, and would, therefore, not alleviate the costs incurred to comply with source-country taxation.[170] In addition, for fixed amount thresholds, a mechanism of indexation for inflation should be foreseen to avoid the erosion of the threshold's effectiveness.[171] In the eventuality where the virtual PE threshold is not incorporated in the international tax law by means of a multilateral instrument but added to DTTs via bilateral negotiations, countries should aim for being consistent with respect to the fixed amount thresholds adopted in the different treaties that they revise (or enter into).[172]

The revenue threshold, in the form of a fixed amount or a ratio, should be determined with great care, notably in the light of a detailed analysis of the data market. A bar set too high or too low would compromise the efficiency of the new nexus. On this point, establishing a ratio relating to the overall wealth of a jurisdiction may prove to be more challenging than determining a fixed amount.

8.3.2.3.3 User Threshold

Inclusion of a user threshold would indeed ensure to cover only the enterprises that effectively have a significant digital presence in the territory of the source country. For the purposes of this threshold, location of users needs to be determined. Such determination could be made in an efficient manner by establishing the number of devices employed to access a platform through which the digital services are supplied.[173]

As tempting as it may sound from a policy perspective, establishing a threshold on the basis of the time spent by users on the platform would be a genuinely challenging endeavour. Number of users can be determined with ease by means of IP addresses of users' devices and, alternatively, similar methods of geolocation.[174] Determining the time spent per user on the platform requires, however, a more detailed supervision and recording of users' activities, which may reveal to be

169 Arnold (2003), p. 487.
170 Arnold (2003), p. 487.
171 Arnold (2003), p. 488.
172 Arnold (2003), p. 488.
173 See article 4(4) of the Proposed Directive.
174 As indicated in article 4(6) of the Proposed Directive.

problematic from both a technical and a data protection perspective. Even more importantly, the relevant amount of time necessary to form a nexus is likely to vary from one multi-sided platform to another, which would result in the establishment of a variety of sets of rules for different platforms. For instance, time spent on a search engine is necessarily shorter than the time spent on a social media platform. Accordingly, it would not be reasonable to use the time threshold determined for the latter in the context of the former. It would, thereupon, be more suitable to base the threshold on number of users.

Whether such threshold should consist of a fixed number of users or a percentage of the overall population forms a subject of discussion. A high fixed number threshold may cause the non-formation of a PE in countries with smaller populations, and, by contrast, a low fixed number threshold may cause the formation of a PE even in the absence of a significant digital presence in countries with larger populations. For instance, a rather low threshold of 25,000 users may be considered as 'significant' from the perspective of Switzerland, while it would correspond to a relatively small group of users in India. Using a certain *percentage of the general population* as a threshold may, thus, constitute a more viable option. The exact amount of that percentage should be determined on the basis of a thorough analysis, taking into consideration the relevant economic and social factors. Needless to state, such percentages could also be established by means of bilateral negotiations.

The key issue in determining the user threshold is not to opt for a percentage (or a fixed number) too high as to prevent the formation of a PE in the majority of the cases or too low as to cover even the most trivial business activities. The threshold chosen should be apt to reflect the existence, or the absence, of a significant digital presence in the source country.

From a practical perspective, it would be more appropriate to determine *a user threshold per year*, and not per month, to avoid the need to take into consideration minor fluctuations of number of users occurring throughout a year, especially in cases where the threshold is kept low. In line with this approach, it would be advisable to refer to the aggregate number of users in the context of a fixed number threshold and to an average percentage in the context of a percentage threshold.

8.3.2.3.4 Business Contract Threshold

Number of business contracts concluded for the supply of digital services may also constitute a relevant threshold,[175] either in combination with a user threshold or as an alternative threshold. For the threshold to reflect the significant presence as closely as possible, should be taken into consideration all business contracts concluded

[175] See also article 4(3)(c) of the Proposed Directive. The business contract threshold suggested here deviates from the one suggested by the Proposed Directive, which limits the relevant business contracts to the ones concluded between customers located in the source country and the non-resident multi-sided platform.

between all customers regardless of their location and the non-resident multi-sided platform in relation to users located in the source country. The overall number of business contracts would allow to make an accurate assessment of the commercial activity conducted by multi-sided platforms with respect to a source country's user base.

Like the user threshold, the business contract threshold should be designed to measure the intensity of the digital presence as efficiently as possible. The relevant number of contracts should be determined with great caution and a *per annum* approach is advisable to be able to take into consideration the overall business activities of the platform. For instance, the Proposed Directive sets the threshold to 3000 business contracts for the supply of digital services in the relevant tax period.[176]

8.3.2.3.5 Conclusion

Establishing a virtual PE nexus constitutes the most efficient remedy to the current inability to submit multi-sided platforms to corporate income tax in the countries in relation to which such platforms are conducting their commercial activities. Annual number of users along with (or as an alternative to) annual number of business contracts are apt to demonstrate the level of intensity of the business activities conducted by a platform in a given country. The fulfilment of such thresholds should be construed as amounting to the 'significant digital presence' of the enterprises concerned in source countries.

8.3.3 *Attribution of Profits*

As well-known in the international tax law practice, the existence of a PE in a source country does not automatically give rise to profits taxable in the country concerned. The presence of a PE and the amount of profits attributable to a PE constitute two separate and distinct issues.[177] On its own, a PE may indicate 'nothing' about how much profit is earned in a country.[178] A multi-sided platform having a virtual PE in a source country may still not be submitted to tax, due to lack of profits, or more

[176] See article 4(3)(c) of the Proposed Directive.

[177] Arnold (2003), p. 479.

[178] Cui (2019), (electronic version).

precisely lack of profits that are attributable to the source country under the internationally accepted standards.

As is common knowledge, for the purposes of profit attribution, the PE and the non-resident enterprise are considered as independent entities transacting at arm's length price, a concept borrowed from article 9 of the OECD and UN Model Conventions, regulating associated enterprises. Arm's length prices can be defined as prices determined in conformity with the arm's length principle, which "[...] requires associated enterprises to charge the same prices, royalties and other fees in relation to a controlled transaction[179] that would be charged by independent parties in an uncontrolled transaction[180] in otherwise comparable circumstances".[181] Profit attribution, therefore, relies on two fictions: the fiction that the PE forms an independent entity from the non-resident company that it forms a part of; and the fiction that these two entities are engaging in actual transactions that should be evaluated at an arm's length price.[182]

Creation of a virtual PE nexus adds a further layer of fiction into the equation. Users and customers of a multi-sided platform are first treated as a part of the platform's business operations to establish the existence of a significant digital presence (*first fiction*). Once the significant digital presence is established and a virtual PE is formed, the PE has to be treated as a separate entity from the multi-sided platform (*second fiction*). Then, it must be controlled whether the transactions between this separate entity and the multi-sided platform are effectuated at an arm's length price (*third fiction*). Most importantly, all this analysis has to be conducted within the context of a business model, which is, by its very definition, based on strategic pricing decisions in relation to its different sides and either complete or partial (like the digital barter) subsidisation of certain sides.[183]

On top of all these 'usual' complexities, the largest multi-sided platform companies as well as the start-ups active in the field are constantly loss making. While making losses forms an almost natural part of the process for the latter, heavily investing in new technologies with a primary aim of safeguarding, or strengthening, the market dominance, 'burning money' in the process and remaining in an almost constant loss-making position constitute a business decision made by the former.[184]

When all these factors are taken into consideration, one has to agree with Cui's observation on the application of the arm's length principle in the context of

[179] The term 'controlled transaction' refers to "[t]ransactions between two enterprises that are associated enterprises with respect to each other" [OECD TP Guidelines (2022), Glossary, p. 21].

[180] The term 'uncontrolled transaction' refers to "[t]ransactions between enterprises that are independent enterprises with respect to each other" [OECD TP Guidelines (2022, Glossary, p. 27].

[181] IBFD (2009), pp. 23–24.

[182] Cui (2019), (electronic version).

[183] On this point, see also Cui (2019), (electronic version).

[184] On this point, see, for example Cui (2019), (electronic version); and Galloway (2017), pp. 31–51 (for the example of Amazon.com). See, also, Partington (2021).

multi-sided business models: it "would generate only gibberish".[185] As already mentioned by a number of contributors, creating a new nexus cannot produce an efficient solution in the context of the digital economy, unless an effective mechanism for international profit allocation can be established.[186] The remaining issue is, thus, determining an attribution method that could make a *quasi*-meaningful sentence out of the gibberish.

To this end, the Proposed Directive opts for the use of the profit split method unless the taxpayer proves that an alternative method based on internationally accepted principles is more appropriate with regard to the results of the functional analysis.[187] The splitting factors include expenses incurred for R&D and marketing as well as number of users and data collected.[188] In a similar vein, Hongler and Pistone opt for the profit split method, with one modification: an upfront allocation of a part of the overall income to source jurisdictions.[189] Petruzzi and Buriak favour a net-basis taxation with an emphasis on the activities conducted by users and the value of data.[190]

In our opinion, instead of persisting in the attempt to make a meaningful sentence out of the gibberish, another language, which is meaningful to begin with, could be employed. Undoubtedly, this would imply departing from the internationally accepted allocation methods to a certain extent. That being said, the method that we are suggesting can also be considered as internationally accepted, at least by a number of countries.

Implementation of a data collection tax eliminates the need to take into consideration the value of user data in the context of the taxation of digital services provided by multi-sided platforms to their customers. The same observation will hold true when (and if) a data labour market is established. As a matter of fact, within the framework of such a market, users will be properly remunerated for their data. In the context of the virtual PE nexus suggested by the authors, number of users is utilised merely for the purposes of establishing the existence of a significant digital presence. The virtual PE nexus applies, thus, exclusively to the profits obtained by multi-sided platforms from their activities conducted through the side(s) of their businesses that do(es) not involve users. Such activities are effectuated in a digital manner, which implies that the majority of the business contracts relating to them are concluded in a digital manner. Digitally conducted activities leaving digital traces; the gross revenues obtained from the activities covered by the virtual PE nexus can, therefore, be established with ease. This has also been demonstrated by the current practice with DSTs.

[185] Cui (2019), (electronic version).

[186] See, for example, Olbert and Spengel (2017), p. 19; and Petruzzi and Buriak (2018), p. 5.

[187] Article 5(6) of the Proposed Directive.

[188] Article 5(6) of the Proposed Directive.

[189] Hongler and Pistone (2015), p. 34.

[190] Petruzzi and Buriak (2018), pp. 18–19.

Once the gross revenues obtained on the basis of the activities rendered to the customers in relation to a source country are determined, a number of deductions should be allowed to determine the net profits. To this end, the amount of data collection tax paid by the platform in the source country would be deducted. Other expenses directly related to the activities rendered in relation to the source country would also be deducted. For instance, should 'streaming fees'[191] effectively enter into force in certain countries, the amount of such fees would be deductible from the taxable base. Marginal costs of providing digital services would also be deducted. As often mentioned, marginal costs relating to digital services tend to be rather low. The 'true' costs of highly digitalised enterprises reside in their fixed costs. Deduction of fixed costs, however, should be examined with caution. Investments made in excessive R&D activities with a primary aim of safeguarding market domination, should not be considered as deductible expenses from a source country point of view. Only the relevant portion of the fixed costs that are directly related to the provision of digital services should be allocated to source countries.

This approach requires a detailed, and more importantly an internationally accepted, guidance on deductions. Such guidance should ideally be prepared under the aegis of the OECD or the UN in order to be able to reach a certain international consensus and a harmonised approach. To ensure that the deduction of the fixed costs can be properly conducted, discussions may be undertaken with the representatives of multi-sided platforms as well. The guidance prepared can be inserted in the OECD Commentary and/or the UN Commentary, along with the insertion of the new virtual PE nexus.[192]

The virtual PE nexus proposed in this book can, thus, function as a 'DST on steroids'. It forms a solution that is compliant with the current system of the DTTs and allows to take into consideration the costs incurred in the course of the business activities. More importantly, it avoids the need to take into consideration the value of user-generated data.

[191] Margrethe Vestager, Executive Vice President of the European Commission of a Europe Fit for the Digital Age, indicated in May 2022, that the tech giants are generating a lot of traffic in the telecommunication networks, which gives rise to the necessity of investing substantial amounts to the telecommunications infrastructure. There is an ongoing discussion in the EU, to establish whether the tech giants should pay 'streaming fees' to cover a portion of those telecommunications costs, which can only go higher with the more frequent use of metaverse. For more information on the issue, see for example Sweney (2022) and Euronews (2022).

[192] Beyond any doubt, reaching an international consensus on the issue can constitute an extremely burdensome endeavour. The solution that can be implemented as an end-result, however, would be efficient and largely devoid of complexity. It is also clear that some modifications in the wording of articles 5 and 7, along with their Commentaries, have to be made in view of accommodating the new nexus. Such modifications are beyond the scope of this book.

8.3.4 'Digital Services Fees' as a Residual Category

The virtual PE solution may be complemented with the incorporation of 'digital services fees' in DTTs as a residual category.[193] The suggestion being made is actually inserting a clause that is akin to article 12B of the UN MC. The purpose of adding such a clause is twofold: (i) it would capture the income earned by highly digitalised enterprises that do not meet the significant digital presence threshold of the virtual PE; and (ii) it may be used as a stand-alone solution by countries who are not willing to implement the virtual PE.

Inserting a digital services fees clause that allows source taxation up to a certain percentage of the gross amount of the payments underlying the income from digital services may constitute an easy-to-implement solution, involving less complexities than the virtual PE. The core issue is to define the concept of digital services fees with precision. On this point, article 12B of the UN MC may provide guidance. Also, it would be advisable to employ the same tax rate that is used in the articles relating to royalties and technical services fees (if existent).

It is important to emphasise that article 12B of the UN MC does not fully address the value generated by data collected through the use of 'free' digital services.[194] In line with the reasoning adopted in this book, however, such value does not need to be addressed in the context of *step three*.

In the eventuality where both a virtual PE and a withholding tax on digital services fees are incorporated into a DTT, highly digitalised enterprises that do not meet the virtual PE threshold may be given a right to opt-in to the net taxation scheme suggested under the PE instead of being submitted to a withholding tax under a specific article regulating digital services fees.[195] Such a right can be implemented with ease in countries making use of the digitalisation in their tax administrations. For instance, a non-resident enterprise opting for being taxed under the virtual PE scheme may be granted a QR Code, which could be used by payers of the digital services fees when they are proceeding with their payments to demonstrate that no withholding needs to be effectuated.[196]

References

Arnold B (2003) Threshold requirements for taxing business profits under tax treaties bulletin tax treaty monitor. Bull Int Taxat IBFD 57(10):476–492

[193] On this point, see also the proposal of a new article regulating 'cyber-based services' made by Zhu [Zhu (2014)].

[194] UN Commentary (2021) ad article 12B, § 10.

[195] In a similar vein, see Brauner and Pistone (2017), p. 6.

[196] In a similar vein, see Brauner and Pistone (2017), p. 6.

Avi-Yonah RS (2018) The international implications of Wayfair. Tax Notes Int, 9 July 2018:161–165

Brauner Y, Pistone P (2017) Adapting current international taxation to new business models: two proposals for the European Union. Bull Int Taxat IBFD 71(12):1–7. (electronic version)

Collin P, Colin N (2013) Task Force on Taxation of the Digital Economy, Report to the Minister for the Economy and Finance, the Minister for Industrial Recovery, the Minister Delegate for the Budget and the Minister Delegate for Small and Medium-Sized Enterprises, Innovation and the Digital Economy

Cui W (2019) The digital services tax: a conceptual defense. Tax Law Rev 73(1):69–11. (electronic version)

Elliffe C (2021) Taxing the digital economy: theory, policy and practice, 1st edn. Cambridge University Press Cambridge Tax Law Series

Escribano E (2019) A preliminary assessment of the EU proposal on significant digital presence: a brave attempt that requires and deserves further analysis. In: Cid JMA, Gutiérrez JAF, González-Barreda PAH (eds) Combating tax avoidance in the EU: harmonization and cooperation in direct taxation, 1st edn. Wolters Kluwer, pp 559–599

Euronews (2022) Should Google, Meta and Netflix help pay for telecoms networks? Why not, says EU's Vestager. Euronews. https://www.euronews.com/next/2022/05/02/should-google-meta-and-netflix-help-pay-for-telecoms-networks-why-not-says-eu-s-vestager. Accessed 7 July 2023

European Commission (2018) COM (2018) 147 final 2018/0072 (CNS), Proposal for a Council Directive Laying Down Rules Relating to the Corporate Taxation of a Significant Digital Presence

Galloway S (2017) The four: the hidden DNA of Amazon, Apple, Facebook, and Google, 1st edn. Penguin Random House UK

Holmes K (2014) International tax policy and double tax treaties: an introduction to principles and application, 2nd edn. IBFD

Hongler P, Pistone P (2015) Blueprints for a new PE nexus to tax business income in the era of the digital economy, IBFD Working Paper 20 January 2015

IBFD (2009) Rogers-Glabush J (ed) IBFD international tax glossary, 6th revised edn. IBFD

Jiménez AM (2018) BEPS, the digital(ized) economy and the taxation of services and royalties. Intertax 46(8/9):620–638

Malherbe J, Maraia J, Traversa E (2013) Article 12. In: Danon R, Gutmann D, Oberson X et al (eds) Modèle de Convention fiscale OCDE concernant le revenu et la fortune, Commentaire, Helbing Lichtenhahn, Editions Francis Lefebvre, pp 471–495

Mehta A (2016) Importance of human involvement for income characterization as fees for technical services. Asia-Pac Tax Bull IBFD 22(2):1–5. (electronic version)

OECD (2015) Addressing the Tax Challenges of the Digital Economy, Action 1 - 2015 Final Report, OECD/G20 Base Erosion and Profit Shifting Project. OECD Publishing

OECD (2019) Model tax convention on income and on capital 2017 (full version). OECD Publishing

OECD (2022) OECD transfer pricing guidelines for multinational enterprises and tax administrations 2022. OECD Publishing

Olbert M, Spengel C (2017) International taxation in the digital economy: challenge accepted? World Tax J IBFD 9(1):2–27. (electronic version)

Partington R (2021) G7 tax reform: what has been agreed and which companies will it affect? The Guardian. https://www.theguardian.com/world/2021/jun/07/g7-tax-reform-what-has-been-agreed-and-which-companies-will-it-affect. Accessed 7 July 2023

Petruzzi R, Buriak S (2018) Addressing the tax challenges of the digitalization of the economy – a possible answer in the proper application of the transfer pricing rules? Bull Int Taxat IBFD 72(4a):1–19. (electronic version)

Pickering A (2012) Enterprise Services: General Report. In: International Fiscal Association Cahiers 2012, Vol. 97a, 1st edn. International Fiscal Association, pp 17–62

Pinto D (2003) E-commerce and source-based income taxation IBFD Doctoral Series 6, 1st edn. IBFD

Rocha SA (2012) Enterprise services: Brazil. In: International fiscal association cahiers 2012 Vol. 97a 1st edn. International Fiscal Association, pp 155–168

Sasseville J, Skaar A A (2009) Is there a permanent establishment? General Report. In: International Fiscal Association Cahiers de Droit Fiscal International Vol. 94a, 1st edn. International Fiscal Association, pp 17–63

Schön W (2018) Ten questions about why and how to tax the digitalized economy. Bull Int Taxat IBFD 72(4/5):278–292

Sweney M (2022) European telecoms chiefs call on tech firms to share internet network costs, The Guardian. https://www.theguardian.com/business/2022/sep/26/european-telecoms-chiefs-call-on-tech-firms-to-share-internet-network-costs. Accessed 7 July 2023

Tadmore N (2008) Royalties (Article 12 OECD Model Convention). In: Lang M, Pistone P, Schuch J et al (eds) Source versus residence: problems arising from the allocation of taxing rights in tax treaty law and possible alternatives, 1st edn. Wolters Kluwer, pp 107–128

UN (2021) United Nations model double tax convention between developed and developing countries 2021. United Nations

van der Bruggen E (2001) Source taxation of consideration for technical services and know-how with particular reference to the treaty policy of China, India and Thailand. Asia-Pacific Tax Bull 7(3):42–60

Zhu Y (2014) Proposed Changes to the UN Model Convention Dealing with the Cyber-based Services. UN Tenth Session of the Committee of Experts on International Cooperation in Tax Matters Doc. No. E/C.18/2014/CRP.9

Chapter 9
Distribution of Data (Step 4)

Abstract This chapter aims to analyse sale of data, transfer of data and granting the right to use a database from an international tax law perspective. To this end, first are analysed sale of data and the grant of the right to use a database, with the primary aim of determining the most appropriate distributive rule(s) applicable to these two transactions. Then, intra-group transfers of raw and processed data are succinctly elucidated, with the main purpose of establishing the most suitable transfer pricing method(s) that could be applied to the transactions concerned.

9.1 Introduction

In the distribution and use phases of the data's life cycle, raw and/or processed data can typically be sold to third parties or transferred to group companies. Data may also be organised in the form of a database and the right to use such database may be granted to third parties. Albeit not forming an exclusive list of potential uses of data, these three circumstances form the most frequent cases of data distribution and utilisation apart from the internal use of data, which is examined in Chap. 8.

Sale of data, transfer of data and granting the right to use a database have been regrouped under *step four* as per the methodology adopted in this book.[1] Taxation modalities of these transactions are analysed in this chapter. First are analysed the sale of data, along with the grant of the right to use a database, to third parties. Then, intra-group transfers of data are briefly examined.

These three transactions are exclusively analysed from an international tax law perspective, since this book opts for a tax policy and international tax law analysis without elaborating details of different domestic laws. In other words, for the purposes of this chapter, it is assumed that a cross-border element exists in each transaction. In any case, to date, no country seems to have implemented a national tax regime with the specific purpose of regulating transactions involving data.[2] In

[1] For more information on the methodology, see Sect. 3.5.

[2] With the notable exception of DSTs. See Sprague (2022), p. 15.

X. Oberson, A. E. Yazıcıoğlu, *Taxation of Big Data*,
https://doi.org/10.1007/978-3-031-43372-6_9

addition, very little judicial or administrative domestic law guidance directly addressing data transfer transactions exists.[3] The same observation holds true from an international tax law perspective. No specific modifications were made in relation to data transactions with the notable exception of article 12B of the UN MC and no specific detailed guidance were issued. The examination in this Chapter is, therefore, conducted on the basis of the existing international tax law rules. On this point, as also noted by Sprague, an international harmonisation on certain issues, perhaps through additions to the OECD MC, would be welcome[4] to avoid divergent interpretations and other complexities in practice.

The purpose of the analysis is to characterise the transactions from a treaty law point of view, and for the intra-group transfers of data to identify the most suitable TP method. The following developments are made from a general and broad perspective, and depending on the specific circumstances of a given case they may not hold true. As in the majority of transactions, a case-by-case basis analysis reveals to be necessary to proceed with the most appropriate characterisation as well as to determine the most suitable TP method.

The general tendencies of countries with respect to data transactions were summarised by Sprague in the General Report prepared on the subject of 'Big data and tax- domestic and international taxation of data driven business' for the International Fiscal Association's 2022 Congress. The observations made in this Chapter regarding such tendencies are based on the General Report.

9.2 Sale of Data and Right to Use a Database

Data brokers and other data aggregators may sell data either in their raw form or after processing. Although multi-sided platforms traditionally employ the data they collect for internal purposes, the possibility of sale of data by multi-sided platforms cannot be excluded altogether.[5] In a similar vein, data brokers, data aggregators and multi-sided platforms may form databases and grant the right to use such databases to third parties. A database does not necessarily entail the existence of processed data; it may also be formed by means of regrouping the raw data under a number of categories.

To analyse the sale of data and the right to use a database from an international tax law perspective, first the income categories of the OECD and UN Model Conventions that may potentially apply to such transactions are determined. Then, each income category is examined one by one to establish the most relevant category on the basis of different circumstances.

---

[3] Sprague (2022), p. 26.

[4] Sprague (2022), p. 39.

[5] See Sect. 3.2 for more information on different actors of the data market.

9.2.1 Determining the Most Relevant Income Categories

To pinpoint the most pertinent income categories, first a number of private law aspects should be examined in detail. Such thorough review falls beyond the scope of this book. The determination of the relevant income categories is based on the following general observations made in the General Report:

- There is no property right in data *per se*, even when data have been aggregated in a database.
- While no copyright protection is offered to data, some jurisdictions (such as the EU and the US) offer copyright protection to databases to a certain extent.
- Some jurisdictions provide for limited *sui generis* protection for collections of valuable data sets.
- Device manufacturers can generally not claim trade secret ownership rights in data collected by the devices they sell to consumers. Consumers typically cannot claim trade secret rights in the data collected by the devices they own either, due to the fact that they cannot normally claim a competitive advantage from keeping such data secret.[6]

In the absence of a property right or an intellectual property right granted specifically to processed data, there is no need to make a distinction between the sale of raw data and the sale of processed data from a tax law point of view. In cases where, a valuable data set, which is likely to be a processed data set, has been granted a *sui generis* protection, the tax treatment of such data set would differ from other data sets. The fact that databases may be protected by copyright in some jurisdictions should be taken into consideration while conducting the tax law analysis. The intellectual property rights are highly relevant for the purposes of this analysis, and such rights are determined by the domestic laws of the contracting states. As a consequence, in this chapter, the tax law analysis is effectuated on the basis of the general assumptions.

In conformity with this preliminary information, the potential income categories that may apply to a sale of data, either in a raw or processed form, and to a grant of the right to use a database are the following: business profits (article 7 of the OECD and UN MCs), royalties (art. 12 of the OECD and UN MCs), technical services fees (article 12A of the UN MC) as well as income from automated digital services (article 12B of the UN MC).[7]

A hierarchy can be construed between these different income categories for the purposes of the analysis at hand. To the extent where income derived from a sale of data or a grant of the right to use a database can be classified as royalties or technical

---

[6]Sprague (2022), p. 19. These aspects, along with other additional features, are also mentioned in Sect. 2.4.1.

[7]These income categories are examined in detail in Sect. 8.2. For the ease of reading, reference will mostly be made only to the articles and not to the Model Conventions for the rest of this chapter (e.g., article 12A instead of article 12A of the UN MC).

services fees, article 12 or article 12A would become applicable. No hierarchy exists between these two categories, notably due to the fact that they give rise to a similar tax treatment, or even to an identical tax treatment[8] if the rates of source taxation designated in both articles are the same. Should article 12 or article 12A not be applicable to a particular case, article 12B may enter into consideration as explicitly indicated in its paragraph 7. In cases where the beneficial owner of the royalties, technical services fees or income from automated digital services carries on business in the country where such income arises through a PE situated therein and the income concerned is effectively connected with the PE, article 7 becomes applicable. Finally, in all other cases, article 7 is applicable as a residual category, since sale of data and granting the right to use a database would, in any case, clearly constitute business profits for data brokers, data aggregators and multi-sided platforms.

9.2.2 *Royalties*

In a nutshell, to give rise to a royalty, a data set or a database should: (i) be protected by an intellectual property right or a similar right; or (ii) amount to a secret formula or process; or (iii) constitute information concerning industrial, commercial or scientific experience; or (iv) form an industrial, commercial or scientific equipment (under the UN MC). Moreover, the transaction should consist of granting the *use or the right to use* the data set or the database. In cases where the transaction involves the transfer of the full ownership of the data set or the database, the payment cannot be considered as being made 'for the use of, or the right to use' the property concerned, and, therefore, it cannot represent a royalty.[9]

9.2.2.1 Intellectual Property Right or a Similar Right

A data set formed by the aggregation and combination of raw or processed data is not granted an intellectual property, or a similar right, protection under most jurisdictions. Accordingly, in the majority of cases, sale of data cannot be considered as giving rise to a royalty payment on the basis of the existence of an intellectual property or a similar right. A valuable data set, which is likely to be a processed data set, may benefit from a *sui generis* protection in certain countries. In the same vein, a database may be protected by copyright. As a consequence, royalties may arise from the use or the right to use of a 'valuable data set' or a database.

[8] To the extent where a source taxation is provided by article 12 of the applicable DTT. The OECD MC does not provide for a source taxation in its article 12. For more information on this point, see Sect. 8.2.2.

[9] OECD Commentary (2019) ad article 12, § 8.2.

In its analysis conducted on software, the OECD Commentary underlines that payments made for acquisition of partial rights in the copyright are considered to represent a royalty, where the consideration is for granting of rights to use the program in a manner that would, without such license, constitute an infringement of copyright.[10] Are enumerated as examples of such arrangements, licenses to reproduce and distribute to the public software incorporating the copyrighted program, or to modify and publicly display the program.[11] The key is to obtain the exploitation of rights that would otherwise be the sole prerogative of the copyright holder.[12] In cases where the rights acquired in relation to software are limited to enable the entity acquiring the rights to operate the program, for instance when a copy of the program is purchased and the acquirer is granted limited rights to reproduce the program, payments obtained in exchange for the rights concerned should be qualified as business profits under article 7.[13] These observations may apply, *mutatis mutandis*, to the use or the right to use of a valuable data set or a database.

In accordance with the OECD Commentaries, it can be argued that characterisation as royalties is appropriate only where the acquirer has been provided with comprehensive rights of use for commercial exploitation of the valuable data set or the database. This would imply, among others, the right to reproduce the data set/database, or its substantial parts, for distribution, either as such or in a modified form; or the right to make the date set/database, or its substantial parts, publicly available for download, either as such or in a modified form. In cases where the acquirer merely obtains the right to employ the database or the data set, i.e., accessing, reading, printing or otherwise exploiting the data in the regular course of its business, the payment cannot be qualified as royalties.[14]

A purchase of a data set or a payment to obtain access to a database would be effectuated, in most cases, with an aim of having access to and making use of the underlying data in the purchaser's own business operations. For instance, an insurance company may obtain from a data broker an access to a database containing health data. In a similar vein, a newspaper can purchase from a data broker the data segments prepared in relation to a certain part of the population. Under such circumstances, the payment made in exchange of data cannot be classified as royalties. Should the purchaser acquire rights going beyond this mere use of the underlying data and amounting to a commercial exploitation of the database or the data set itself, a royalty qualification may be appropriate, depending on the other factual circumstances of the case. Whether a right going beyond the utilisation of data has been granted can be assessed on the basis of a number of factors, such as restrictions imposed on the purchaser to further transfer the data, the amount and the

[10] OECD Commentary (2019) ad article 12, § 13.1.

[11] OECD Commentary (2019) ad article 12, § 13.1.

[12] OECD Commentary (2019) ad article 12, § 13.1.

[13] OECD Commentary (2019) ad article 12, § 14.

[14] On this point see also, the Austrian Report by Daurer et al. (2022), p. 132.

structure of the compensation paid in exchange for the rights as well as the structuring of the data set/database in accordance with the specific needs of the purchaser.[15]

9.2.2.2 Secret Formula or Process and Information Concerning Industrial, Commercial or Scientific Experience

Secret formula or process entails the transfer of a secret technology, a trade secret or, in the context of software, an information supply about the ideas and principles underlying the program, such as logic, algorithms and programming languages or techniques.[16] Information concerning industrial, commercial or scientific experience (know-how) constitutes a similar concept. It refers to information of an industrial, commercial or scientific nature arising from previous experience, which has practical application in the operation of an enterprise and from the disclosure of which an economic benefit can be derived.[17] Transfer of know-how implies, thus, divulging a special knowledge and experience, which remain unrevealed to the public.[18] In this respect, it differs from a provision of services, in which one of the parties undertakes to use the customary skills of its calling to execute the work itself for the other party.[19]

To produce a data set or to elaborate a database, an enterprise has to establish a process, which certainly has an intrinsic value. An algorithm allowing the enterprise to collect the data generated by users and to process them to derive ready-to-use information, combined with other techniques that need to be employed during this phase, would amount to a 'secret process' or 'know-how'. That being said, sale of a data set or granting the right to use a database does not divulge, *per se*, any information about the process employed to prepare the data concerned. No information relating to the underlying process, such as the functioning of the algorithm, the data collection methods and optimisation of the information obtained, is communicated to the purchaser. Consequently, in general, these transactions do not result in a transfer of know-how or a secret formula.

9.2.2.3 Industrial, Commercial or Scientific Equipment

The use or the right to use industrial, commercial or scientific equipment, contained in article 12(3) of the UN MC, refers to cases in which the owner of the equipment

---

[15] See also, Sprague (2022), pp. 30 and 34.

[16] On this point, see OECD Commentary (2019) ad article 12, § 9.1, 9.2, 11.5, 14.3.

[17] OECD Commentary (2019) ad article 12, § 11.

[18] OECD Commentary (2019) ad article 12, § 11.1.

[19] OECD Commentary (2019) ad article 12, § 11.2.

earns profits from letting another person use that equipment.[20] While the term 'equipment' is defined neither in the Model Convention nor in its Commentary, and thus has to be interpreted in accordance with the domestic laws of the contracting states,[21] the main feature that should always be present is the use of the equipment in the performance of a task.[22] The equipment should form a tool used by a business, in the sense that it is not enjoyed for its own sake.[23] For instance, a car rented by a tourist cannot be considered as an equipment, whereas a car used in the course of a business activity, such as for transportation of merchandise, would.[24] Although not expressly mentioned, the reference is being made to physical equipment as it can be deducted by the developments made and the examples used in the Commentary.

As argued by Zhu, an online database could be construed as a type of 'virtual' equipment for its users.[25] In the context of treaties referring to 'industrial, commercial or scientific equipment', the terms 'or online database' could be inserted to allow this distributive rule to encompass databases as well.[26] Nonetheless, such a modification has not been made in the UN MC and the authors are not aware of any DTTs containing such formulation. Albeit considering a database as a virtual equipment may be deemed as a natural consequence of the evolving technology, in the authors' opinion, the scope of the term 'equipment' cannot be extended to cover 'virtual equipment' by the sole means of interpretation.

Consequently, under the current wording of the rule, a sale of data or a grant of the right to use a database cannot be considered as the use of an equipment in the source country.

9.2.2.4 Conclusion

On the basis of the analysis conducted, it can be concluded that a typical sale of data or grant of the right to use a database is not likely to give rise to royalties under article 12.

[20] UN Commentary (2021) ad article 12, § 18.

[21] The need for a domestic law interpretation derives from article 3(2) of the UN MC.

[22] UN Commentary (2021) ad article 12, § 19.

[23] UN Commentary (2021) ad article 12, § 19.

[24] UN Commentary (2021) ad article 12, § 19.

[25] Zhu (2014), p. 11.

[26] Zhu (2014), p. 11.

9.2.3 Technical Services Fees

The term 'technical services fees' is defined as any payment in consideration for any service of a managerial, technical or consultancy nature in article 12A(3) of the UN MC. For a payment to be considered as a technical services fee, the enterprise should perform work for the customer in exchange for that payment, instead of merely transferring intellectual property, know-how or experience.[27] The contours of the technical services fees concept cannot be drawn with certainty. As already explained, the scope of the term may vary largely from DTT to DTT and from domestic law to domestic law.[28] Accordingly, in the absence of a particular case and a precise DTT, only a broad observation regarding sale of data and grant of the right to use a database can be made.

In most cases, granting a right to use a database would not necessitate any 'work' to be conducted for the customer. Such transaction typically involves a mere grant of authorisation to access a ready-to-use database. Subsequently, payments made in exchange for using a database should, under normal circumstances, not be classified as technical services fees.

On the other hand, depending on the particular circumstances of a case, it is highly probable for a sale of data to be considered as a service of a consultancy nature. This holds especially true in cases where a specific service in relation to data is expected from a data broker, such as fraud detection services and identification of the relevant data segments. In a similar vein, a company may be requested to render consultancy services on the basis of the user data that it collects and processes. In accordance with the domestic laws of the contracting states and the applicable DTT, the qualification as a service of a technical nature may also enter into consideration.

Despite not being applicable to the entire transaction, the characterisation of technical services fees may also be appropriate for services accompanying the right to use a database or sale of a data set. For instance, a data broker may provide further assistance once the right to use a database is granted or a certain data is sold. Even if the income deriving from the right to use the database/sale of data is considered as royalties (or is otherwise characterised) in such a case, the assistance services that accompany that right may be categorised as technical services fees. In that eventuality of 'mixed' transactions, the payment obtained in exchange for the bundle of services may be disaggregated and separate allocation rules can be applied to each distinct part, unless one of the components is only of an ancillary nature. In that latter case, the allocation rule applicable to the predominant part of the transaction applies to the entire transaction.[29]

---

[27] UN Commentary (2021) ad article 12A, § 4.

[28] For more information on this point, see Sect. 8.2.3.

[29] UN Commentary (2021) ad article 12A, § 4; see also OECD Commentary (2019) ad article 12, § 11.6.

It can, therefore, be concluded that in the context of DTTs containing a specific allocation rule on technical services fees, payments obtained from sale of data can fall within the scope of application of the rule concerned.

9.2.4 Income from Automated Digital Services

Article 12B of the UN MC specifically mentions 'supply of user data' as an automated digital service.[30] The UN Commentary defines 'the supply of user data' as "the provision of data to a third-party customer in respect of users of a digital interface, which is collected, compiled, aggregated or otherwise processed into data through an automated algorithm".[31] The Commentary also underlines that the source of data is not relevant *per se*, they may be collected as raw data by the enterprise or acquired from another business.[32]

In line with this definition, should a remuneration obtained in exchange for sale of data or the right to use a database cannot be characterised as royalties or technical services fees, it can, *a priori*, be qualified as an income from automated digital services. Since the definition provided by the Commentary refers to collection, compilation, aggregation or otherwise processing; raw data, which have to be collected and compiled, are also covered by the provision concerned. Accordingly, there is no need to make a distinction between sale of raw data and processed data.

Nevertheless, for article 12B to be applicable, the service giving rise to the income should satisfy the definition given in paragraph 5. As per the definition concerned, the concept of automated digital services refers to any service provided on the Internet or another electronic network, in either case requiring minimal human involvement from the service provider.[33] The key issue is, thus, to determine whether the human involvement necessary to render the service could be considered as minimal or not.

As already mentioned, the UN Commentary specifies that a service is considered to be automated "[. . .] when the user is able to make use of the service because of equipment and systems being in place, which allow the user to obtain the service automatically, as opposed to requiring a bespoke interaction with the supplier to provide the service".[34] Only the engagement from the side of the supplier is pertinent in determining whether a service requires minimal human involvement.[35] The efforts put by the purchaser to utilise the automated system in view of obtaining a

[30] Article 12B(6)(b).

[31] UN Commentary (2021) ad article 12B, § 58(b).

[32] UN Commentary (2021) ad article 12B, § 58(b).

[33] Article 12B(5).

[34] UN Commentary (2021) ad article 12B, § 53.

[35] UN Commentary (2021) ad article 12B, § 53.

customised result is not relevant.[36] In addition, the human intervention necessary to create, support or maintain the system needed for the provision of services, or making other generic adjustments that are unrelated to individual purchaser requests do not constitute a human involvement within the meaning of the definition of the term.[37] The same holds true for human intervention where the provision of services to new purchasers involves very limited human response to individual purchaser requests/input.[38] The purpose is, thus, to cover the services in relation to which enterprises can scale up and provide the same type of services to new purchasers with minimal human involvement.[39] This is typically the case of services for the provision of which the marginal costs incurred in terms of additional human involvement is almost nil.[40] Consequently, a service is considered to be an 'automated digital service', to the extent where, once developed, it can be provided to one purchaser, or to many more, on an automated basis with the same basic business process.[41] On the contrary, a service is considered to be a 'non-automated digital service', in cases where its provision requires a proportionate increase in the costs per unit in connection with providing the services to new customers.[42]

On the basis of these explanations, it would appear that granting the right to use a database may form an automated digital service, to the extent where it does not fall within the scope of application of articles 12 and 12A. In fact, once the database is prepared and ready-to-use, access to it may be granted to one or a number of purchasers on an automated basis with the same basic business process. The same may hold true in the context of sale of data, so long as the sale concerned is not combined with a provision of service, which would require a specific additional work to be conducted per purchaser. For instance, letting a purchaser access a database and utilise the fraud detection tool regularly updated by the data broker constitutes an automated digital service. In a similar vein, purchasing from a data broker a list of aliases commonly used to commit fraud amounts to an automated digital service. Requesting the data broker to run a check on a list of individuals in its own fraud detection system, on the other hand, involves an additional work by the data broker and cannot amount to an automated digital service.

---

[36] UN Commentary (2021) ad article 12B, § 53.
[37] UN Commentary (2021) ad article 12B, § 53.
[38] UN Commentary (2021) ad article 12B, § 53.
[39] UN Commentary (2021) ad article 12B, § 54.
[40] UN Commentary (2021) ad article 12B, § 54.
[41] UN Commentary (2021) ad article 12B, § 54.
[42] UN Commentary (2021) ad article 12B, § 54.

9.2.5 *Business Profits*

Should the income derived from the sale of a data set or the grant of a right to use a database cannot be qualified as a royalty, technical services fees or income from automated digital services, it would generally fall within the category of business profits. Such an outcome would be quite common in practice, since the majority of DTTs, along with the OECD MC, do not contain specific distributive rules relating to technical services fees and income from automated digital services. The characterisation of business profits would also apply in cases where the royalties, technical services fees or income from automated digital services obtained are closely connected to the PE that the data broker, the data aggregator or the multi-sided platform has in the source country.

The concept of PE is examined in detail in Chap. 8,[43] and no further examination reveals to be necessary for the purposes of sale of data and granting the right to use a database. Two cases specifically related to data collection and aggregation activities should, however, be examined, since they may give rise to the creation of a PE in source countries during the conduct of activities directly related to data. The first case is the collection of data by means of a software installed in equipment utilised by users and the second case consists of the activities conducted through a data centre situated in the source country. Since no specific guidelines were issued in relation to activities relating to data, the analysis is to be conducted on the basis of the general rules and commentaries, or the commentaries made in relation to similar cases, such as activities conducted through a server.

9.2.5.1 Devices Equipped with Software and Connected to Internet

Devices equipped with software and connected to internet (IoT) are collecting data on a continuous basis. While the physical device is owned by the user, the non-resident enterprise may remotely access to the device through the software. It has, therefore, to be determined whether such devices may form a PE of the non-resident enterprise. The fixed place PE being the only type of PE that could be created with regard to devices collecting data, the analysis is limited to this type of PE.

Nothing excludes a device collecting data from being characterised as a fixed place of business under article 5. In fact, equipment and machinery may form 'fixed places' and they are expressly included in the description of the concept of PE.[44] The fact that devices can be moved does not imply that they cannot be considered as 'fixed' either. The criterion of being 'fixed' does not refer to the possibility of moving the equipment, but to the fact that it is actually moved.[45]

---

[43] See Sect. 8.2.1.

[44] See OECD Commentary (2019) ad article 5, § 41.

[45] OECD Commentary (2019) ad article 5, § 125.

To form a PE, the equipment should be at the disposal of the non-resident enterprise. It can be argued that since the enterprise can access the devices remotely by virtue of the software incorporated in them, such devices should be deemed as being at the disposal of the enterprise. An interesting ruling was rendered in Denmark with regard to this aspect. In relation to devices equipped with a software that could be activated and deactivated by the owner of the device, the Danish Tax Assessment Board asserted that such devices could not be considered as being at the disposal of the non-resident enterprise. According to the Board, it should be deemed that the devices are at the disposal of their owners, who could choose to activate or disable the software imbedded in them.[46] As a matter of fact, if the owner of the device opts for never activating the software or for deactivating it, the non-resident enterprise cannot access the device; and, therefore, this latter cannot be at the disposal of the former. This line of reasoning raises, however, an interesting question. Should the relevant criterion be the possibility of deactivation or being actually deactivated—or never activated—? In the authors' opinion, the reasoning with regard to transportability of the movable objects should be applied in this context as well. The criterion of being 'at the disposal' should not be determined on the basis of the possibility of deactivation, but on the basis of the fact that the software is effectively deactivated, or not activated.[47]

A key issue, which would become relevant in the majority of cases, is to determine whether the collection of data occurring through devices can be considered as forming a part of the core business of the non-resident enterprise. Beyond any doubt, this examination needs to be effectuated on a case-by-case basis in accordance with the primary business activities conducted by the enterprise concerned. For instance, from the perspective of a data broker, it is clear that data collection forms a part of the core business activities. On the other hand, for a computer manufacturer that collects data through the computers it manufactured in view of conducting a performance analysis to proceed with the necessary ameliorations in the manufacturing process, collection of data seems to occupy an auxiliary role in the sense of article 5(4).[48] As a matter of fact, collection of data would generally reveal to be an auxiliary activity in cases where a company's value chain is predominantly physical and data analytics are used as a tool to improve efficiency or reduce costs.[49]

Devices equipped with software are generally not commercialised by data brokers. They are manufactured and sold by enterprises active in a variety of sectors, usually with a predominantly physical value chain. Accordingly, in most cases data collected by such devices would not constitute a part of the core business of the non-resident enterprise, and a fixed place PE would not be formed. One notable

---

[46] Ruling from the Danish Tax Assessment Board, SKM 2019.643.SR; cited by the Danish Reporters, Bundgaard and Kjærsgaard (2022), pp. 315–317.

[47] The same line of reasoning is adopted with regard to the 'fixed' character of the place of business in the context of servers, see OECD Commentary (2019) ad article 5, §125.

[48] On this point, see also Olbert and Spengel (2017), p. 17.

[49] Schwarz (2022), p. 787.

exception to this general observation is multi-sided platforms, a side of which is built on users. Albeit being dependent on the factual circumstances, the data collection effectuated by such multi-sided platforms via devices is likely to constitute a part of their core business. Under these circumstances, the devices concerned could give rise to a fixed place PE.

9.2.5.2 Data Centres

Data centres may also form a fixed place PE in the country where they are situated. As in the case of servers, when they are directly controlled by the non-resident enterprise, which would be a rather rare occurrence, data centres could potentially form a fixed place PE.[50] In contrast, when they are controlled by a third-party, data centres are not likely to constitute a fixed place PE. The most problematic determination issues would occur in cases where the data centre assets are owned by an entity related to the non-resident enterprise. This observation is based on the necessity for the fixed place of business to be *at the disposal* of the non-resident enterprise for the creation of a PE.

While no formal legal right to use a particular place is required for a place to constitute a fixed place PE, the enterprise should have the effective power to use that location and should establish a certain degree of presence at that location.[51] Such an effective power to use a location would not be deemed to exist in cases where the enterprise's presence at the location concerned is so intermittent and incidental that the location cannot be considered as a place of business of the enterprise.[52] For instance, in cases where the employees of an enterprise have access to the premises of associated enterprises, which they often visit without, however, working in these premises for an extended period of time, no PE can be formed.[53]

This requirement of an effective power to use a location constitutes the reason for which data centres employed by non-resident enterprises but owned by third-parties cannot form a PE of the former. On this point, the reasoning followed by the OECD Commentary with regard to servers is unequivocal and may be followed, *mutatis mutandis*, for the purposes of data centres as well. As is common knowledge, servers are usually maintained by Internet service providers. The enterprise owning the website pays a fee to the service provider in exchange for the hosting of the website on a server at the disposal of this latter. According to the OECD Commentary, these agreements do generally not result in the server and its location being at the disposal of the enterprise commercialising online products.[54] This holds true even in cases

[50] Assuming that all the conditions necessary for the constitution of a PE is fulfilled, including the necessity to conduct a part of the core activities through the data centre.

[51] OECD Commentary (2019) ad article 5, §12.

[52] OECD Commentary (2019) ad article 5, §12.

[53] OECD Commentary (2019) ad article 5, §12.

[54] OECD Commentary (2019) ad article 5, §124.

where the enterprise is able to determine the particular server on which the website should be hosted.[55] As a consequence, in cases where the data centre is owned by a third-party, a fixed place PE is not likely to be formed.

In cases where the data centre is owned by an entity related to the non-resident enterprise, this latter may have a certain power to use the centre concerned. For the location to be considered at its disposal, however, the employees of the non-resident enterprise should not only have access to the premises but also effectively work in the data centre for a certain period of time, and not merely visit it from time to time. This point can be illustrated by a ruling rendered by the Danish Tax Assessment Board. In determining whether a data centre situated in Denmark and owned by the Danish subsidiary of a foreign parent company could be considered as a PE of the non-resident parent company, the Board took into consideration the following factors, among others: the parent company and a small group of its employees were granted the permission to visit the data centre, only if accompanied by the employees of the Danish subsidiary; the parent company's own employees—located outside Denmark—monitored the efficiency of the hardware and software installed in the data centre, installed and uninstalled applications, carried out maintenance of the hosted applications and handled the software and data stored on the servers; in cases where a server would not work properly, it could be shut down using this remote access; in the opinion of the Board such remote access would not differ from the standard terms seen in any cloud computing arrangement.[56] Under these, and similar, circumstances, it cannot be argued that the data centre is at the disposal of the non-resident enterprise.

Consequently, the existence of a data centre in a country may provoke the formation of a PE of a non-resident enterprise; depending notably on the effective control exercised by the enterprise on the centre concerned, along with the relevant factual circumstances of the case.

9.3 Intra-Group Transfers of Data

TP relating to data transactions has not been subject to specific regulations or guidelines from an international tax law perspective. The situation seems to be the same in domestic laws. According to the General Report, there is little or no precedent in the national jurisprudence applying TP principles to transactions involving big data.[57]

Data collected by different group companies can be shared within the group in a variety of ways. Data may be collected by one entity and shared in their raw form

[55] OECD Commentary (2019) ad article 5, §124.

[56] Ruling from the Danish Tax Assessment Board, SKM2016.188.SR; cited by the Danish Reporters, Bundgaard and Kjærsgaard (2022), pp. 317–318.

[57] Sprague (2022), p. 43.

with another entity, which anonymises and sends them to another entity, which produces data segments and submits them to a separate entity, etc. Data collected, processed and otherwise analysed by the group can be made available to all entities through a database, or be stored by one particular group company. In short, no specific model of data exploitation exists. The purpose of this section is not to examine in detail all potential data transfers within a group from a TP perspective. One must admit that such a task would reveal to be extremely challenging.

This section focuses only on two types of intra-group data transfers that can occur: (i) transfer of raw data between associated enterprises,[58] and (ii) transfer of processed data between associated enterprises. The examination made in this section is built on the shoulders of the analysis previously effectuated in this book. It is important to remind that both raw data and processed data constitute intangibles;[59] and that both raw data and processed data have value, despite the fact that the latter is more valuable.[60] Determining the exact value of data, or a data set, is an extremely cumbersome endeavour, and, under certain circumstances, even determining an approximate value could constitute a significant challenge.[61] The general tendency seems to consist of considering that the value of processed data is rather high, while the value of raw data is trivial, or even non-existent. On this point, it must be borne in mind that intangibles that do not have book value in the company's balance sheet may be taken into consideration for the purposes of the TP analysis, since they may actually have considerable value under the actual economic circumstances.[62] Raw data are one of such intangibles.

9.3.1 Transfer of Raw Data

A group company may collect user data in the country where it is situated, store such data, form a raw data set and send it to another company (or companies) of the same group situated in another country (or countries). These collection and storage phases represent lower costs and risks when compared to data processing and analytics phases. The end-product, set of raw data, is also of a lower value.

A transfer of raw data can typically occur in two different scenarios. *First*, the user data may be collected by means of the software embedded in devices or equipment employed by consumers, temporarily stored, compiled to form a raw data set and transferred to another group company in order to be analysed and to be utilised in ameliorating the performance of the devices concerned. Such a data set does not have a market value *per se*. As a matter of fact, that particular data are only

[58] Article 9 of the OECD and UN MCs.

[59] See Sect. 2.4.

[60] See Sect. 3.4.1.

[61] See Sect. 4.5.

[62] Gómez (2017), p. 286.

valuable to the group manufacturing the devices. They may also reveal to be valuable for the competitors of the group concerned, but are likely to be safeguarded as a trade secret and not sold to such competitors. As a consequence, it can be stated that the intangibles in question have particular characteristics and are typically transferred only among associated enterprises. Accordingly, reliable and potentially comparable transactions cannot be found with ease.[63] In these particular circumstances, in the authors' opinion, the most suitable TP method would be cost plus.[64] The entity collecting, storing and aggregating—at least to a certain extent—the data has to incur costs, which can be established in a rather objective manner. The ratio to be applied to the costs concerned can also be determined on a rather objective basis, by conducting an analysis of different sectors. It is clear that the amount obtained through the use of the cost plus method cannot reflect the actual value of the raw data set. That being said, the actual price of such a raw data set cannot be established through the employment of the other TP methods either. Should the application of the cost plus method not be appropriate due to the specific circumstances of the case at hand, a second method that could be utilised would be the transactional profit split method, which would, however, require a much complex analysis when compared to cost plus.[65]

Second, the local user data may be collected by a subsidiary of a data broker, a data aggregator or a multi-sided platform, temporarily stored, compiled in the form of a raw data set and transferred to another group company in order to be analysed and employed in rendering services to customers. Under this scenario, the transaction involves a rather easily identifiable service, which is frequently bought and sold in the data market. In fact, raw data sets are commercialised between data brokers, as well as between other data aggregators and data brokers.[66] It becomes, therefore, possible to identify identical or comparable services traded in the market, the prices and terms of which may be publicly reported. As a consequence, in the authors' opinion, the most suitable method to evaluate the transactions undertaken in this second scenario would be the comparable uncontrolled price method.[67]

---

[63] OECD TP Guidelines (2022), § 6.138.

[64] On this point, see also OECD TP Guidelines (2022), § 6.143. For further information regarding the cost plus method, see OECD TP Guidelines (2022), § 2.45–2.61.

[65] For further information regarding the transactional profit split method, see OECD TP Guidelines (2022), § 2.114–2.183.

[66] On this point, see Sect. 3.2.2.

[67] For further information regarding the comparable uncontrolled price method, see OECD TP Guidelines (2022), § 2.14–2.26.

9.3.2 *Transfer of Processed Data*

The 'true' value of data is unearthed at the phases of processing and analytics. What is meant by processing may vary from enterprise to enterprise and the end-results that may be obtained through data analytics differ to a great extent. Each data set may present intrinsic characteristics, rendering the establishment of reliable comparable transactions quite cumbersome. Domination of the processed data market by a few companies adds an additional layer of difficulty.[68] Specificities of MNEs combined with a market dominated by a limited number of 'siren servers' render the TP analysis extremely challenging to say the least.

Given the complexities of the data market and the challenges inherent to the application of the TP methods, the most suitable method for the purposes of processed data transfers seems to be the transactional profit split method. On the basis of different and accurate splitting factors, such as expenses incurred for research, development and marketing, number of users as well as location of the qualified personnel, a more or less accurate analysis can be conducted.

References

Bundgaard J, Kjærsgaard LF (2022) Denmark. In: International fiscal association 2022 Berlin congress, cahiers de droit fiscal international: big data and tax-domestic and international taxation of data driven business, vol 106B. International Fiscal Association, pp 303–320

Daurer V, Kofler G, Mayr G (2022) Austria. In: International fiscal association 2022 Berlin congress, cahiers de droit fiscal international: big data and tax-domestic and international taxation of data driven business, vol 106B. International Fiscal Association, pp 127–144

Gómez OG (2017) Transfer pricing of intangibles in the digital economy. In: Kerschner I, Somare M (eds) Taxation in a global digital economy: Schriftenreihe IStR Band 107, 1st edn. Linde Verlag Ges.m.b.H, pp 283–300

OECD (2019) Model tax convention on income and on capital 2017 (full version). OECD Publishing

OECD (2022) OECD transfer pricing guidelines for multinational enterprises and tax administrations 2022. OECD Publishing

Olbert M, Spengel C (2017) International taxation in the digital economy: challenge accepted? World Tax J IBFD 9(1):2–27. (electronic version)

Schwarz AM (2022) Switzerland. In: International Fiscal Association 2022 Berlin Congress, Cahiers de droit fiscal international: big data and tax-domestic and international taxation of data driven business, Vol. 106B. International Fiscal Association, pp 769–794

Sejati U (2017) Value creation in the digital economy. In: Kerschner I, Somare M (eds) taxation in a global digital economy: Schriftenreihe IStR Band 107, 1st edn. Linde Verlag Ges.m.b.H, pp 257–282

Sprague GD (2022) General Report. In: International Fiscal Association 2022 Berlin Congress, Cahiers de droit fiscal international: big data and tax-domestic and international taxation of data driven business, vol 106B. International Fiscal Association, pp 11–71

[68] On this point, see also Sejati (2017), pp. 275–276. For further information on market dominance, see Sect. 3.3.3.

UN (2021) United Nations model double tax convention between developed and developing countries 2021. United Nations

Zhu Y (2014) Proposed Changes to the UN Model Convention Dealing with the Cyber-based Services. UN Tenth Session of the Committee of Experts on International Cooperation in Tax Matters Doc. No. E/C.18/2014/CRP.9

Chapter 10
VAT on Data Transactions

Abstract This chapter aims to analyse data transactions from a value added tax perspective. To this end, first are examined characterisation and place of location of data supplies, along with the concept of fixed establishment. Then, a more detailed analysis is conducted on data supply as a taxable event. The EU VAT Directive is chosen as a model in this exercise.

10.1 General Explanations

10.1.1 Main Principles

As a consumption tax, VAT is levied on all supplies of goods and services, including digital services, by domestic enterprises, subject to a certain threshold requirement. Following the destination principle, the import of goods or services should also be subject to VAT. It follows that, digital transactions, including transactions with data, should be included in the scope of existing consumption tax regimes to ensure the neutrality of these taxes.[1] Online domestic sellers of goods and services that use digital platforms for domestic consumers are also subject to VAT and required to register. By contrast, non-resident enterprises, which offer those services to consumers in the domestic market, notably through online platforms, are not always included in the scope of the VAT regime.

For the purposes of the analysis conducted in this chapter, the EU VAT regime is used as a model. The EU VAT regime has the merit of presenting a well-developed system, which also serves as a model to other countries in the modification and the application of their domestic VAT legislations. Using a model system is important for the purposes of this chapter, since it allows conducting a more thorough examination of the subject matter.

---

[1] Aslam and Shah (2020), p. 48.

X. Oberson, A. E. Yazıcıoğlu, *Taxation of Big Data*,
https://doi.org/10.1007/978-3-031-43372-6_10

In the EU, VAT is governed by the EU VAT Directive.[2] This Directive has been modified many times since its adoption.[3] Until recently, following the Directive of 19 October 1992,[4] the Member States had to apply a standard rate which may not be less than 10%. One or two reduced rates, which may not be less than 5%, could apply to specific categories of goods and services. On 5 April 2022, the EU Council adopted a new Directive which includes an overhaul of VAT rates.[5] This text represents a compromise which, in addition to the standard rate of 15%, opens possibilities to apply two reduced rates of 5% to goods and services in 24 categories, in accordance with a revised Annex III of the VAT Directive. That list includes digital services, such as internet access and live streaming of cultural and sports events. This new Directive will become applicable as of 1 January 2025.

Another modification pertinent to the digital economy was effectuated with regard to online platforms. As of 1 January 2021, under the so-called 'e-commerce VAT package', online platforms allowing the selling of goods to EU customers face new obligations. Online platforms that facilitate imports to customers goods (B2C) of a value below EUR 150 are deemed to receive the supplies from the initial seller (deemed B2B supply) and to sell the goods to the final customer (deemed B2C supply).[6] Therefore, online platforms are liable for the collection of VAT on the supplies to the final consumers and the remittance of the VAT to the competent tax authorities. Determining the place of location and the applicable VAT rate is also incumbent upon the platforms concerned.

10.1.2 Characterisation and Place of Location of Data Supplies

In general sense, *transactions with data* are characterised as a supply of services. It follows that, under the EU VAT regime, data transactions with business customers (B2B) are governed, in accordance with the general rules for services, by the destination principle (reverse charge system). The characterisation as a service applies both for the electronic transfer of raw data, the transfer of aggregated or structured data, or the grant of a right to use copyrights in a database.[7] As a

[2]Council Directive 2006/112/EC of 28 November 2006 on the common system of value added tax (hereinafter 'the VAT Directive').

[3]For an overview see, for example, Ben Terra and Kajus (2023), pp. 269–273.

[4]Council Directive 92/77/EEC of 19 October 1992 supplementing the common system of value added tax and amending Directive 77/388/EEC.

[5]Council Directive 2022/542 of 5 April 2022 amending Directives 2006/112/EC and (EU) 2020/285 as regards rates of value added tax.

[6]Bal (2020).

[7]Pinkernell (2022), p. 383.

consequence, the recipient of the services has to charge VAT and can credit the input VAT on its supply subject to tax (output VAT).

In the EU, under the so-called e-commerce VAT package, remote digital service providers are also required to register to VAT on supplies offered to resident customers in the place of consumption. Services involving data supplied to private customers (B2C) are also governed by the destination principle, following the special rule for electronic services and telecommunication.

An exception to the general characterisation as a service could occur in cases where the data are stored in a physical disk, which is then transferred to the end user of the data. In this particular case, the transaction could be characterised as a supply of goods, in a similar manner as the sale of standardised software stored on physical media.[8]

A crucial issue with regards to characterisation occurs in the context of co-renting a space within a data centre. Should such co-renting be categorised as a supply of goods, namely a transfer of power of disposition of a tangible good, or in a more general manner, a service? This issue has already been examined in a number of court decisions. For instance, in the A Oy case,[9] the CJEU had to consider the case of a Finnish enterprise, offering co-renting arrangements to non-resident enterprises. The Finnish tax administration characterised these activities as a supply of goods, subject to VAT in Finland, because of their link with the infrastructures situated in Finland. The CJEU did not follow that reasoning. According to the Court, while the supply consists of putting the infrastructure (the data centre along with ancillary activities, such as supervision, electric supply and cleaning) at the disposal of the customer, it does not include the right to dispose of the infrastructure as an owner. A similar reasoning was followed by a French Court in the case of co-renting arrangements supplied by the French enterprise Equinix France to non-resident companies by means of its four data centres located in France.[10]

10.1.3 The Presence of a Fixed Establishment

A significant issue is to determine whether the presence of an online platform could trigger the presence of a fixed establishment (FE) for VAT purposes. As is common knowledge, the concept of FE forms a key aspect in this area, since it determines the place of supply for services and also establishes the right to deduct input VAT.[11] The CJEU traditionally emphasised, by means of a long series of cases starting from the leading Berkholz case, the need for a physical presence in order for a place of

[8] Pinkernell (2022), p. 383.

[9] Case C-215/19, *A Oy* of 02 July 2020.

[10] Administrative Court of Appeal of Versailles, Case n. 19VE00526, *SASU Equinix France* of 08 December 2020.

[11] De la Feira (2021), pp. 201–205.

business to qualify as a FE. In the Welmory case, the Court started to depart from the well-established requirement of a physical presence.[12] Recent case law, notably Titanium, seems, however, to deviate from this 'new' approach because the Court considered that without human resources, a property could not be regarded as a FE.[13] The development of this case law is briefly summarised below to serve as a basis for our analysis.[14]

In the Berkholz case, the CJEU ruled that the installation of gaming machines on board of a sea ship sailing on the high seas outside the national territory could be regarded as a FE "only if the establishment entails the permanent presence of both the human and technical resources necessary for the provision of those services and it is not appropriate to deem those services to have been provided at the place where the supplier has established his business".[15] Following the same approach in the ARO Lease case, the CJEU held that a leasing company "which does not possess in a Member State either its own staff or a structure which has a sufficient degree of permanence to provide a framework in which agreements may be drawn up or management decisions taken, and thus enable the services in question to be supplied on an independent basis",[16] cannot be considered having a FE in that country.

In the DFDS A/S case,[17] the main issue was to determine the VAT liability of a Danish company (namely DFDS A/S) in the UK, in respect of package tours sold on its behalf by its English subsidiary (namely DFDS Ltd.). In particular, the Court had to determine whether a travel agent operating in another Member State through a subsidiary could be considered as independent or not. For the purposes of the case concerned, the CJEU held that where a tour operator established in one Member State (in the case at hand Denmark) provides services to travellers through the intermediary of a company operating as an agent in another Member State (for the purposes of the case at hand the UK), VAT is payable on those services in the latter country if that company, which acts as a mere auxiliary organ of the tour operator, has the human and technical resources typical of a FE. The legal ownership of the premises in which the business activities take place, on the other hand, is irrelevant.[18]

This jurisprudence was confirmed in the Pflanzer case,[19] in which the Court reminded that "the term fixed establishment implies a minimum degree of stability derived from the permanent presence of both the human and technical resources necessary for the provision of the given services [. . .] It thus requires a sufficient degree of permanence and a structure adequate, in terms of human and technical

[12] De la Feira (2021), p. 204.

[13] Case C-931/19, *Titanium Ltd.* of 03 June 2021.

[14] For a summary, see also De la Feira (2021), pp. 201–205; and Oberson (2019), pp. 98–104.

[15] Case C-168/84, *Gunter Berkholz* of 04 July 1985, § 19.

[16] Case C-190/95, *ARO Lease BV* of 17 July 1997, § 19.

[17] Case C-260/95, *DFDS A/S* of 20 February 1997.

[18] Cannas (2016), p. 99.

[19] Case C-73/06, *Planzer Luxembourg Sàrl* of 28 June 2007.

resources, to supply the services in question on an independent basis".[20] In the field of transport activity, that term entails the existence of "at least an office in which contracts may be drawn up and daily management decisions taken, and a place where the vehicles used for the said activities are stored".[21]

Later, in the Welmory case,[22] the Court had to deal with a *receiving* FE. Welmory Ltd., established in Cyprus, organised online auction sales by selling packages of 'bids', which incorporated the right to make an offer to purchase goods at a higher price than the price last offered. Welmory Ltd. had a cooperation agreement with a Polish company, under which it agreed to provide the latter with the services of making available an internet auction site with a domain name, including the supply of associated services related to the leasing of the servers and the display of the goods to be auctioned. The Polish company agreed to sell goods on that website. The issue was whether the services were supplied to Cyprus or a FE of the Polish company in Poland, thus subject to VAT at a standard rate of 22%.

With respect to the case, the Court ruled that, in order to be considered as having a FE, the Cypriot company should have in Poland "at the very least a structure characterized by a sufficient degree of permanence, suitable in terms of human and technical resources to enable it to receive in Poland the services supplied to it by the Polish company and to use them for its business, namely running the electronic auction system in question and issuing and selling bids".[23] The Court also took account of article 11 of Council Implementing Regulation 282/2011[24] and considered that such definition was directly inspired by the case law of the CJEU.[25] In addition, the Court added that the fact "that the business carried out by the Cypriot company, consisting in operating a system of electronic auctions which comprises, first making an auction website available to the Polish company and, secondly, issuing and selling 'bids' to customers in Poland *can be carried on without requiring an effective human and material structure in Polish territory* is not determinative. Despite its character, such a business requires at least a structure that is appropriate in terms especially of human and technical resources, such as appropriate computer equipment, servers and software".[26] In other words, the Court ruled that the Cypriot company which received services supplied by the Polish entity must be regarded as having a FE in Poland, only if it had a sufficient degree of permanence and suitable structure in terms of human and technical resources to enable it to receive the services supplied to it and use them for business.[27]

---

[20] Case C-73/06, *Planzer Luxembourg Sàrl* of 28 June 2007, § 54.

[21] Case C-73/06, *Planzer Luxembourg Sàrl* of 28 June 2007, § 55.

[22] Case C-605/12, *Welmory* of 16 October 2014.

[23] Case C-605/12, *Welmory* of 16 October 2014, § 59.

[24] Council Implementing Regulation (EU) No 282/2011 of 15 March 2011 laying down implementing measures for Directive 2006/112/EC on the common system of value added tax.

[25] Case C-605/12, *Welmory* of 16 October 2014, § 58.

[26] Case C-605/12, *Welmory* of 16 October 2014, § 60. Emphasis added by the authors.

[27] Oberson (2019), p. 102.

The WebMindLicences (hereinafter WML) case pertains to the operation of a website offering audio-visual services.[28] In this case, WML, a company registered in Hungary, acquired, in a free of charge manner, from Hypodest Patent Development, a company in Portugal, know-how enabling a website to allow individuals throughout the world to take part in real-time erotic interactive audio-visual services. On the same day, WML made that know-how available, by a license agreement, to Lalib, a company in Portugal.

The competent tax authorities argued that the transfer of WML's know-how to Lalib did not correspond to a genuine economic transaction, as the know-how was exploited by WML. In line with this reasoning, the authorities held that the exploitation that was subject to VAT had taken place on Hungarian territory. In particular, the CJEU found that: "It is incumbent upon the referring court to analyse all the circumstances of the main proceedings in order to determine whether that agreement constituted a wholly artificial arrangement concealing the fact that the services at issue were not actually supplied by the company acquiring the license but were in fact supplied by the company granting it. It should examine whether the establishment of the place of business or fixed establishment of the company acquiring the license was not genuine, whether that company, did not possess an appropriate structure in terms of premises and human and technical resources and whether it did not engage in that economic activity in its own name and on its own behalf, under its own responsibility and at its own risk".[29]

It can be argued that in Welmory, the Court had revisited the notion of a FE and seemed to depart from the 'strict physicality of the concept'.[30] However, more recent case law tends to deviate from the Welmory case and seems to 'go back' to a more classical notion of a FE. Indeed, in Titanium, the Court was asked whether a property might be characterised as a FE. The request was made with an aim of determining whether Titanium Ltd., established in Jersey and active in property and housing and accommodation management, was subject to VAT on rental income relating to a property located in Austria. The Austrian authorities considered that Titanium had a FE in the country concerned. In its decision, the CJEU summarised the case law as follows: "The concept of 'fixed establishment', in accordance with the Court's settled case-law, implies a minimum degree of stability derived from the permanent presence of both the human and technical resources necessary for the provision of given services. It thus requires a sufficient degree of permanence and a structure adequate, in terms of human and technical resources, to supply the services in question on an independent basis [. . .] A structure without its own staff cannot fall within the scope of the concept of a 'fixed establishment'".[31] Therefore, the Court considered that a property, which is let in a Member State in the circumstance where

[28]Case C-419/14, *WebMindLicenses Kft.* of 17 December 2015.

[29]Case C-419/14, *WebMindLicenses Kft.* of 17 December 2015, § 50.

[30]De la Feira (2021), p. 204.

[31]Case C-931/19, *Titanium Ltd.* of 03 June 2021, § 42.

the owner of that property does not have the infrastructure or their own staff to perform services relating to the letting, does not constitute a FE.

It may be argued that the case law of the CJEU has been transposed in article 11 of the Council Implementing Regulation 282/2011.[32] Indeed, according to article 11(1) of the Regulation, a *receiving* FE, within the meaning of article 44 of the VAT Directive "shall be any establishment characterised by a sufficient degree of permanence and a suitable structure in terms of humans and technical resources to enable it to receive and use the services supplied to it for its own needs". This corresponds to the definition given in the Welmory case. In addition, a *providing* FE, according to article 11(2) of the Regulation, "shall be any establishment [. . .] [c]haracterised by a sufficient degree in permanence and a suitable structure in terms of human and technical resources to enable it to provide the services which it supplies". It is to be noted that the CJEU referred to article 11 of the Regulation in the Titanium case as well.

As De la Feira points out, the concept of FE has been progressively eroded by the globalisation and digitalisation of the economy.[33] From this perspective, the apparent step back to a more traditional definition of the FE concept taken by the CJEU can be regarded as moving in the wrong direction. It has to be acknowledged, however, that the Court is not in a position to simply abandon the rules of article 11 of the Regulation, which also focuses on the concept of physical presence. It follows that despite the recent doors opened by the Welmory case, the traditional approach remains applicable and in the absence of a physical presence in a Member State, an online platform would not qualify as a FE for VAT purposes. In a way similar to the development of the rules applicable to the digital economy for corporate income tax purposes, the concept of FE should also be revised and adapted to the new business models of the digital economy.

10.2 Data Supply as a Taxable Event?

As already mentioned in detail, within the framework of the digital economy, in a first stage, multi-sided platforms offer various online services via information technology (IT) or other telecommunication tools that may be utilised, generally for 'free' or in exchange for a negligible price, by users. The platforms are, then, in a second stage, realising profits through the processing of those data in view of

[32] On this point, see the arguments put forward by van Norden, who considers that the additional criterion of «use for its own needs» is not justified [van Norden (2011), p. 47]. It is to be noted that the contributor expressed this point of view in 2011; i.e., before the *Welmory* decision, rendered in 2014.

[33] De la Feira (2021), p. 205.

rendering advertising or other similar services to their customers or the employment of such data for other commercial purposes.[34]

From a VAT perspective, sale of data by service providers, or any other relevant transfer of services, against consideration to third-party enterprises (which would correspond to the second stage, as identified above) is clearly characterised as a supply of service for consideration, which is subject to VAT in accordance with the destination principle. As a rule, the enterprise recipient of the data may then credit the VAT as input tax.

The first stage, however, i.e., the data 'offered' by users, gives rise to an interesting problem from a VAT point of view. It has to be determined whether provision of data should be regarded as the counterpart of the services provided by the multi-sided platforms. This issue has already been raised by some commentators, who argue that the services provided by social media or search engine platforms are in fact 'paid' by users through provision of data (the digital barter).[35] As already mentioned,[36] the digital barter entails that the employment of digital services offers an economic value to users, who are, in turn, willing to grant service providers access to their data.[37] The digital barter consists, thus, of an exchange of supplies. The issue is especially relevant in B2C transactions in which the enterprises subject to VAT would, in theory, have to charge VAT on their digital supplies offered to consumers beneficiaries. Indeed, in B2B transactions, the potential VAT would correspond to an amount of zero since the beneficiary would be allowed to claim an input VAT credit for the VAT paid.[38]

In any event, in order to be subject to VAT, the transaction should be regarded as a supply of service linked with a consideration corresponding to the value attributable to the data offered in exchange. In other words, under this approach, data collection could be regarded as a taxable supply and VAT should be levied on the deemed value of the data.[39] A debate on this issue exists in Germany (ignited by two articles published in 2015),[40] Austria,[41] Switzerland[42] and most particularly in Italy. In fact, it was announced on 22 February 2023 that the Italian authorities initiated an investigation against Meta, on the basis of the existence of a commercial value of user data collected by the group's platforms (like Facebook) in consideration of allowing users to access the platforms. As per the alleged view of the authorities,

---

[34] For more information on this point, see Sect. 3.2.

[35] Melan and Wecke (2015) pp. 2815–2816; see also Aslam and Shah (2020), p. 50.

[36] For more information on the digital barter, see Sect. 3.5.1.

[37] Schmalenbach-Gesellschaft "Transfer Pricing" Working Group (2020), p. 495.

[38] Schmalenbach-Gesellschaft "Transfer Pricing" Working Group (2020), p. 495.

[39] Zawodsky (2018), p. 606; Marian (2022), p. 560.

[40] By two members of the tax authorities, Melan and Wecke, who were not acting in their official capacities [Pinkernell (2022), p. 381]. See Melan and Wecke (2015).

[41] Daurer et al. (2022), p. 141.

[42] Hug (2020), p. 7.

despite the fact that no fee is charged for registration to such platforms, the value of the user data collected would justify levy of VAT.[43]

Under the general EU VAT rules, supplies of goods and services for consideration are subject to VAT. In other words, there should be a (in)direct and immediate link between the consideration paid and the taxed activity. The characterisation of the 'counterpart' is crucial because, in the absence of a link with the supply received, the overall transaction will qualify as a donation not subject to VAT.[44] In barter transactions, according to the CJEU, the taxable amount should correspond to the subjective value of services or goods received as a counterpart as agreed by the parties to the contract,[45] and not to the value estimated according to objective criteria.[46] Furthermore, for barter transactions, the CJEU, based on the subjective dimension of the taxable base, takes into consideration the value that the supplying entity considers as the counterpart in the exchange.[47] Such a value corresponds to its expenses for the supply of the transaction (following an objective methodology).[48]

In addition, there must be a legal relationship between the person receiving the payment and the person paying it. This specific condition has been confirmed by the CJEU in the Tolsma case.[49] In the case concerned, a musician playing on the public highway, without a contractually stipulated consideration, was not considered as supplying a service within the scope of VAT. Indeed, absent a legal agreement between the listeners, who could give a donation if they choose to do so, and the musician, there was "no necessary link between the musical service and the payments to which it gives rise".[50] It seems that a rather similar reasoning was followed by a German Court, confronted with a potential barter characterisation of a transaction involving the use of data and corresponding services, in the absence of a legal contract between the parties. As pointed out by Pinkernel, this case dates back to 2008 and could be analysed differently, as in the current view, supply of services tends to be considered as a barter transaction under private law in cases where the

---

[43] Parodi (2023).

[44] Lamensch (2021), p. 71.

[45] Case C-230/87, *Naturally Yours Cosmetics Ltd.* of 23 November 1988.

[46] Lamensch (2021), p. 71; Case C-40/09, *Astra Zeneca UK Ltd.* of 29 July 2010, § 28; C-153/12, *Sani treyd EOOD* of 21 March 2013.

[47] Englisch (2021), § 17.251; Case C-33/93, *Empire Stores Ltd.* of 02 June 1994, § 19; Case C-380/99, *Bertelsmann AG* of 03 July 2001, § 38.

[48] On this point, see also Englisch, who argues that this approach focuses too directly on the expenses of the supplying entity. In his view, from a consumption tax standpoint, the concrete expenses of the beneficiary of the supply should be relevant because these expenses, and not those of the supplying enterprise, are indicative of a potential economic ability [Englisch (2021), § 17.251].

[49] Case C-16/93, *R. J. Tolsma* of 03 March 1994.

[50] Case C-16/93, *R. J. Tolsma* of 03 March 1994.

consumer has to consent to the processing of their data to receive the service for 'free'.[51]

It is important to underline that there is a controversy with regards to the 'digital barter' in the VAT literature. Some commentators argue that when users agree to the terms and conditions offered by the platform, including the use of data, the parties conclude a barter transaction, under which the service provider supplies a service 'for free' and, in exchange, the users 'pay' by generating data.[52] By contrast, other commentators argue that no barter transactions exist because the transfer of data, from the perspective of the user, does not form a consideration for the service. According to these commentators, admitting the existence of barter in this particular scenario would go against the main aim of the VAT as a consumption tax.[53]

The EU VAT Committee, following questions from the German Ministry of Finance, published on 3 November 2018, some guidelines on the application of the VAT Directive to barter transactions between platform operators and Internet users.[54] In a nutshell, from a VAT perspective, in the context of the digital barter two services are being exchanged: provision of data and provision of digital services. Whether these services are subject to VAT should be analysed separately on the basis of the applicable articles of the VAT Directive. The conclusions reached by the Committee can be summarised as follows:[55]

1. *Provision of data*: When an individual opts for using a digital service offered without a monetary consideration, which entails agreeing to the terms and conditions of the provider including the grant of the permission to use the individual's data, the individual does not intend to carry out an economic activity and does not deploy the means inherent to such activity.[56] Subsequently, provision of data by users of a digital service offered without monetary compensation cannot be qualified as a taxable supply of services, due to the lack of an economic activity conducted by the former in that particular context.
2. *Provision of digital services*: The provision of a digital service without requiring any monetary consideration but allowing the service provider to collect the data of users, cannot be considered as forming a taxable transaction for VAT purposes, since there is no direct link between the service provided and the consideration received. On this point, the Committee underlines that user data vary largely in quantity and quality from one user to the other, and even include false data. However, to the extent where a sufficient direct link can be established between the digital services provided and the user data collected, the provision of the digital services would be qualified as a taxable transaction. The taxable amount

[51] Pinkernell (2022), p. 382.

[52] Pfeiffer (2016), p. 158; Daurer et al. (2022), p. 141.

[53] Aigner et al. (2017); Daurer et al. (2022), p. 141.

[54] EU VAT Committee (2018).

[55] EU VAT Committee (2018), p. 8.

[56] The requirement of an economic activity is established in article 9(1) of the VAT Directive.

would correspond to the costs incurred by the supplier for the provision of the services concerned to the users.

It appears that, so far, the predominant view is that the link between the service provided by the platform and the non-monetary payment (data) is not sufficient. This view is supported by several arguments. It is, for instance, stated that users have universal access to service providers, while the amount and the quality of the data generated may vary from user to user.[57] In addition, it is argued, the value of the right to use the data, determined from a user perspective, should amount to 'zero', since the user has the possibility to pay for services that protect their identity.[58] Under the case law of the CJEU, data collected as a consideration should have a value from the perspective of the service provider, i.e., on a subjective basis. Some commentators argue that the raw data, as such, represent a very limited value; the value-added potential being realised at a later stage, namely through the analysis, and transformation stages effectuated by service providers ('data mining').[59] This approach places the added value to the data analysis stage, undertaken by service providers, and not to the data generation stage, executed by users.[60]

This majority opinion has been, however, criticised by other commentators who argue, based on the case law of the CJEU, that the conditions of a supply are met because the digital services are, as a rule, rendered exclusively when the right to use the data is granted.[61] In their view, the later use of the data involving third parties, notably for advertising purposes, is distinct from this transaction and should be analysed separately.

In the authors' opinion,[62] a barter transaction exists in all cases where the user data are collected by the service provider and there is a contractual relationship between the two parties.[63] As a matter of fact, in such cases the provider supplies its services with the intention to collect data from the user (subjective criterion). Absent such data, which form the raw material of the services rendered, the provider would, in fact, not even be able to supply services. To our knowledge, supplies of raw material used by taxable enterprises are subject to VAT.

Before concluding, it is important to emphasise that, absent a contractual relationship, a barter transaction would not exist in cases where the user does not consent to the use of their data. This situation would rarely occur in the context of multi-sided platforms and digital service providers, since in most cases general terms of service would be imposed to users by the entities concerned. Data brokers and other

---

[57] In this sense, Aslam and Shah (2020), p. 50.

[58] Schmalenbach-Gesellschaft "Transfer Pricing" Working Group (2020), p. 495.

[59] Aigner et al. (2017), p. 352.

[60] Aigner et al. (2017), p. 353.

[61] Pfeiffer and Ehrke-Rabel (2017), p. 533.

[62] As expressed in detail in Chap. 3.

[63] For a similar point of view, see also Hug (2020), p. 9. It is to be noted that acceptance of general terms of use, and other similar terms and conditions, indicates the existence of a contractual relationship between the parties.

data aggregators may, on the other hand, collect user data without their knowledge and consent.

References

Aigner D, Bräumann P, Kofler G et al. (2017) Digitale Leistungen ohne Geldzahlungen im Internet. 92 SWK 2017. https://www.jku.at/fileadmin/gruppen/150/Team/Georg_Kofler/Aufsaetze_in_Fachzeitschriften/swk-2017-06_0349-0357_191.pdf. Accessed 07 July 2023

Aslam A, Shah A (2020) Tec(h)tonic shifts: taxing the "digital economy". IMF Staff Discussion Note, IMF

Bal A (2020) Online marketplaces and EU VAT: global reach but compliance still local. Kluwer International Tax Blog. http://kluwertaxblog.com/2020/02/26/online-marketplaces-and-eu-vat-global-reach-but-compliance-still-local/. Accessed 7 July 2023

Ben Terra JM, Kajus J (2023) A guide to the European VAT directives 2023. IBFD

Cannas F (2016) The VAT treatment of cloud computing: legal issues and practical difficulties. World J VAT/GST Law 5(2):92–110

Daurer V, Kofler G, Mayr G (2022) Austria. In: International fiscal association 2022 Berlin congress, cahiers de droit fiscal international: big data and tax-domestic and international taxation of data driven business, Vol. 106B. International Fiscal Association, pp 127–144

De la Feira R (2021) On the evolving VAT concept of fixed establishment. EC Tax Rev 30(5/6): 2021–2206

Englisch J (2021) Umsatzsteuer (Kap. 17). In: Seer R, Hey J, Englisch J, Hennrichs J (eds) Tipke/Lang Steuerrecht. 24th edn. Verlag Dr. Otto Schmidt, pp 1037–1268

EU Value Added Tax Committee (Article 398 of Directive 2006/112/EC) (2018) Working Paper No 958. Question concerning the application of EU VAT provisions. taxud.c.1(2018) 6248826-EN. 30 October 2018

Hug T (2020) Daten als mehrwertseuerliches Entgelt? https://www.zsis.ch/artikel/daten-als-mehrwertsteuerliches-entgelt. Accessed 7 July 2023

Lamensch M (2021) Value creation and VAT. In: Haslehner W, Lamensch M (eds) Taxation and value creation. EATLP International Tax Series, IBFD, pp 63–88

Marian O (2022) Taxing data. Brigham Young Univ Law Rev 47(2):511–576

Melan N, Wecke B (2015) Einzelfragen der Umsatzsteuerpflicht "kostenloser" Internetdienste und Smartphone-Apps. DStR 2015:2811–2817

Oberson X (2019) Taxing robots: helping the economy to adapt to the use of artificial intelligence, 1st edn. Edward Elgar Publishing

Parodi E (2023) Italy pursues Facebook-owner Meta for $925 million in sales taxes, Reuters. https://www.reuters.com/technology/italy-pursues-facebooks-meta-925-million-sales-taxes-2023-02-22/. Accessed 7 July 2023

Pfeiffer S (2016) VAT on "free" electronic services? Int VAT Monit 27(3):158–164

Pfeiffer S, Ehrke-Rabel T (2017) Umsatzsteuerbarer Leistungsaustausch durch "entgeltlose" digitale Dienstleistungen. SWK 92(10):532–539

Pinkernell R (2022) Germany. In: International fiscal association 2022 Berlin congress, cahiers de droit fiscal international: big data and tax-domestic and international taxation of data driven business, Vol. 106B. International Fiscal Association, pp 367–384

Schmalenbach-Gesellschaft "Transfer Pricing" Working Group (2020) Data and information as taxable assets. Eur Taxat IBFD 60(11):489–498

van Norden GJ (2011) The allocation of taxing rights to fixed establishments in European VAT legislation. In: van Arendonk H, Jansen S, van der Paardt R (eds) VAT in an EU and international perspective: essays in honour of Han Kogels. IBFD, pp 35–50

Zawodsky F (2018) Value added taxation in the digital economy. Br Tax Rev 5:606–627

Chapter 11
Conclusion: End of the Road?

Throughout this book the authors aimed for suggesting data-specific solutions to data-driven problems. In the midst of a number of critical issues and a great deal of uncertainty, the authors attempted to provide easy-to-implement and efficient proposals on the basis of an interdisciplinary analysis. The core idea of this book is to segregate the utilisation of data into four different yet interdependent steps and construct the tax law analysis on top of these four corner stones. While the issues arising in each step are best solved by means of a distinct solution, each solution is necessarily connected to the others and builds on the previous one(s).

First, it was established that data, the youngest intangible, constitute the raw material of the fourth industrial revolution. *Second*, it was observed that, in its current functioning, the data market is rising on the shoulders of a large number of users and microworkers, while being reigned by a small number of multi-sided platforms. *Third*, it was established that the creation of a data labour market combined with pay-for-privacy options would constitute the best solution to compensate users for the data they generate, but such a system cannot be implemented, at least not yet, for both legal and technical reasons. *Fourth*, the authors suggested a 'second best solution' to adequately compensate users: a data collection tax. This tax would be levied at the stage of collection of data by enterprises and recognise data as a raw material used with a lucrative purpose. *Fifth*, it was argued that the microwork phenomenon is not apt to create new jobs and efficiently tackle the technological unemployment problem. The authors underlined that a part of the funds generated by means of the data collection tax could be used by countries to implement some measures in relation to the unemployment, which is expected to demonstrate an upward trend in the near future. Albeit representing an interesting approach, the idea of a 'robo-data tax' was considered as premature at this stage in time. Such a tax could, however, become necessary, should an economic dependency of certain countries to others begin to occur due to the manners in which user data are collected and processed. *Sixth*, the authors proposed the creation of a virtual PE nexus, combined with a new distributive rule on 'digital services fees', to remedy to the lack of source taxation arising in the context of the digital economy. *Seventh*, the

X. Oberson, A. E. Yazıcıoğlu, *Taxation of Big Data*,
https://doi.org/10.1007/978-3-031-43372-6_11

authors gave their opinion on the most appropriate distributive rules applicable to sale of data and grant of the right to use a database and identified the most suitable TP methods applicable to transfer of data. The scientific analysis concluded with the authors' take on the applicable VAT regime to data transactions. In particular, the authors recognised that, subject to certain conditions, the 'free' digital services provided to users may be deemed as a taxable transaction.

The authors' analysis concludes here, but the discussion on taxation of data does not. The discussion will continue, take different directions on the basis of new technological and legal developments, temporarily solve all the problems and then resume due to new issues arising in the data sector. This never-ending character is perfectly natural when put into the context. The discussion on taxation of data can never end, since it can do nothing but to follow its subject matter, 'data', which can be used, re-used, repurposed and further used and for which the road never ends.

The manufacturer's authorised representative in the EU is Springer Nature Customer Service Centre GmbH, Europaplatz 3, 69115 Heidelberg, Germany. If you have any concerns regarding our products, please contact ProductSafety@springernature.com

Printed and bound by CPI Group (UK) Ltd, Croydon, CR0 4YY
01/10/2026
02225902-0001